An Integrated Approach for Added-Value Products from Lignocellulosic Biorefineries

Alírio Egídio Rodrigues
Paula Cristina de Oliveira Rodrigues Pinto
Maria Filomena Barreiro
Carina Andreia Esteves da Costa
Maria Inês Ferreira da Mota • Isabel Fernandes

An Integrated Approach for Added-Value Products from Lignocellulosic Biorefineries

Vanillin, Syringaldehyde, Polyphenols and Polyurethane

 Springer

Alírio Egídio Rodrigues
Associate Laboratory LSRE-LCM -
Laboratory of Separation and Reaction
Engineering - Laboratory of Catalysis and
Materials, Department of Chemical
Engineering, Faculty of Engineering
University of Porto
Porto, Portugal

Maria Filomena Barreiro
Associate Laboratory LSRE-LCM -
Laboratory of Separation and Reaction
Engineering - Laboratory of Catalysis and
Materials
IPB - Bragança Polytechnic Institute
Bragança, Portugal

Maria Inês Ferreira da Mota
RAIZ- Forest and Paper Research Institute
Aveiro, Portugal

Paula Cristina de Oliveira Rodrigues Pinto
RAIZ- Forest and Paper Research Institute
Aveiro, Portugal

Carina Andreia Esteves da Costa
Associate Laboratory LSRE-LCM -
Laboratory of Separation and Reaction
Engineering - Laboratory of Catalysis and
Materials, Department of Chemical
Engineering, Faculty of Engineering
University of Porto
Porto, Portugal

Isabel Fernandes
Associate Laboratory LSRE-LCM -
Laboratory of Separation and Reaction
Engineering - Laboratory of Catalysis and
Materials
IPB - Bragança Polytechnic Institute
Bragança, Portugal

ISBN 978-3-030-07588-0   ISBN 978-3-319-99313-3   (eBook)
https://doi.org/10.1007/978-3-319-99313-3

This Springer imprint is published by the registered company Springer Nature Switzerland AG
The registered company address is: Gewerbestrasse 11, 6330 Cham, Switzerland

Preface

This book is a result of more than 25 years of research on lignin valorization at the Laboratory of Separation and Reaction Engineering (LSRE), now Associate Laboratory LSRE-LCM at the Department of Chemical Engineering, Faculty of Engineering of the University of Porto (FEUP).

It all started with an invitation of Roberto Cunningham, coordinator of Subprogram IV (Biomass as source of chemicals and energy) of CYTED (Iberian-American Program on Science and Technology for Development), to join project IV.2 "Transformation of lignin on high-added value products" directed by Alberto Venica. The project involved partners from Spain, Portugal, and Latin-American countries.

My task was to produce vanillin from lignin by oxidation and another partner, Mary Lopretti from Uruguay, was using a biochemical route to get vanillin out of lignin. This was in 1990. By chance a Brazilian student applied for a PhD – Álvaro Luiz Mathias from UFPR in Curitiba. He took the subject and started lignin oxidation studies in a Büchi batch reactor. He was more trained in analytical chemistry, and we went to the pulp mill in Cacia of Portucel processing *Pinus pinaster* at that time (now it belongs to The Navigator Company and processes only *Eucalyptus globulus*), took samples of black liquor, and isolated lignin for the oxidation studies aiming at producing vanillin. Alvaro Mathias defended his PhD in 1993 "Production of vanillin from lignin: kinetics and process study."

At that time, I got a Human Capital and Mobility grant from European Union allowing me to hire some postdocs, and Claire Fargues from Nancy (now at IUT d'Orsay) joined this research program. From this research some publications appeared: Fargues et al. 1996 (Kinetics of vanillin production from kraft lignin oxidation, *Ind Eng Chem* Res 35, 28–36); Mathias et al. 1995 (Chemical and biological oxidation of *Pinus pinaster* lignin for the production of vanillin, *J Chem Tech Biotechnol* 64, 225–234); Mathias and Rodrigues 1995 (Production of vanillin by oxidation of pine kraft lignins with oxygen, *Holzforschung* 49, 273–278).

The next step was to make the lignin oxidation process continuous, and a laboratory reactor was built using a structured packing from Sulzer. This was done in the PhD thesis of Daniel Araújo "Development of a process for vanillin production from kraft black liquor" (2008) and published in *Catalysis Today* 147, S330–S335 (2009); *Catalysis Today* 105, 574–581 (2005); and *ChERD* 88, 1024–1032 (2010).

In the meantime, the idea of an integrated process to produce vanillin from the black liquor was being tackled step by step. First using model compounds membrane separation of lignin/vanillin mixture was studied by Miriam Zabkova (now at VULM, Slovakia) and Eduardo Borges da Silva (now at Novozymes, Curitiba, Brazil) and then ion exchange to convert low molecular weight sodium vanillate in vanillin. The whole integrated process was published in *ChERD* 87, 1276–1292 (2009) where degraded lignin was also used to produce polyurethanes. An integrated process to produce vanillin and lignin-based polyurethanes from kraft lignin is sketched below. The contribution to the field of polyurethane materials in LSRE-LCM is being conducted by Prof. Filomena Barreiro, presently at the Polytechnic Institute of Bragança. This was the topic of the PhD of Carolina Cateto (2008), now at ExxonMobil (Belgium).

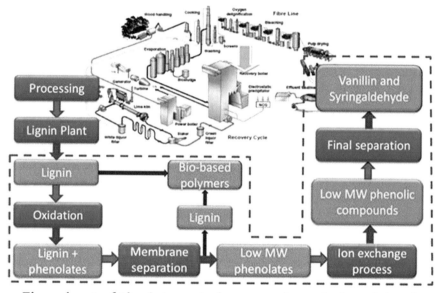

Flow sheet of the integrated process in an industrial unit

In 2009, Dr. Paula Pinto joined the laboratory LSRE and was engaged in various biorefinery projects with industry: Bioblocks with The Navigator Company and another project with a USA company both in lignin topic, and BIIPP (Biorrefinaria Integrada na Industria de Pasta e Papel) in polyphenols from eucalyptus bark. During her stay at LSRE two PhD theses were concluded: Carina Costa "Vanillin and syringaldehyde from side streams of pulp & paper industries and biorefineries" (2017) and Inês Mota "Fractionation of syringaldehyde and vanillin from oxidation of lignin" (2017).

During this period various advances were achieved to produce vanillin and syringaldehyde from lignin. A radar tool (Costa et al. 2015, "Radar tool for lignin classification on the perspective of its valorization," *Ind Eng Chem Res* 54 (31), 7580–7590) was developed to characterize and screen lignins for the production of vanillin and syringaldehyde. Detailed studies on membrane fractionation and adsorption/desorption were done in view of the final objective: Costa et al. 2018 ("Lignin fractionation from *E. globulus* kraft liquor by ultrafiltration in a three stage membrane sequence," *Sep. Pur. Tech* 192, 140–151), Pinto et al. 2017 ("Separation and recovery of polyphenols and carbohydrates from *Eucalyptus* bark extract by ultrafiltration/diafiltration and adsorption processes," *Sep Pur Tech* 183, 96–105), Mota et al. 2016 ("Successful recovery and concentration of vanillin and syringaldehyde onto a polymeric adsorbent with ethanol/water solution," *Chem Eng J* 294, 73–82), and Pinto et al. 2016 ("Performance of side-streams from eucalyptus processing as sources of polysaccharides and lignins by kraft delignification," *Ind Eng Chem Res* 55 (2), 516–526).

Paula Pinto left LSRE-LCM and joined the R&D center RAIZ of The Navigator Company in 2016 followed by Inês Mota in 2017.

The work now continues with three PhD students. Elson Gomes started in 2015; his thesis looking at fractionation of membrane permeate by adsorption separating acids, aldehydes, and ketones (Gomes et al. 2018, "Fractionation of acids, ketones and aldehydes from alkaline lignin oxidation solution with SP 700 resin," *Sep Pur Tech* 194, 256–264) and the final step (crystallization) of the integrated process; Filipa Casimiro hired in 2016 is dealing with the oxidation of species resulting from lignin oxidation, and Carlos Vega-Aguilar (Costa Rica) started his PhD in 2017 on the production of dicarboxylic acids from lignin.

The book is organized in five chapters. Chapter 1 deals with "Chemical pulp mills as biorefineries," providing an overview of delignification industrial processes, integration of new biorefining processes in pulp industry, characterization and classification of lignins using a radar tool, and bark composition and products from bark. Chapter 2 addresses an "Integrated process for vanillin and syringaldehyde production from kraft lignin." It starts with lignin oxidation in batch reactor and then in a continuous structured packed bed reactor. Separation processes (membrane, ion exchange, and adsorption/desorption) are detailed next to separate the low molecular weight phenolics from the degraded lignin and recovery. Chapter 3 is about "Polyurethanes from recovered and depolymerized lignins." After an overview of strategies and opportunities, the use of lignin as such is discussed and then the lignin use after chemical modification is analyzed in particular the case of production of polyurethanes. Chapter 4 deals with "Polyphenols from bark of *Eucalyptus globulus*" including composition of polar extracts and extraction of polyphenols followed by fractionation of ethanolic extracts. Chapter 5 presents "Conclusions and future perspectives."

Porto, Portugal   Alírio Egídio Rodrigues
March 1, 2018

Acknowledgments

We would like to acknowledge financial support along the years from FCT (Foundation for Science and Technology), CYTED (Science and Technology for Development), QREN (Framework for National Strategy), and CRUP/CNRS (Council of Rectors of Portuguese Universities/Centre National Recherche Scientifique).

Contributions from other PhDs and post-docs (Álvaro Mathias, Claire Fargues, Daniel Araújo, P. Sridhar, Carolina Cateto, and Elson Gomes) are also acknowledged.

This work was financially supported by: Project POCI-01-0145-FEDER-006984, Associate Laboratory LSRE-LCM, funded by FEDER through COMPETE2020 – Programa Operacional Competitividade e Internacionalização (POCI) – and by national funds through FCT – Fundação para a Ciência e a Tecnologia.

Contents

Chapter 1
Chemical Pulp Mills as Biorefineries

Abstract Forest-based biorefinery is a multiplatform unit converting lignocellulosic biomass into several intermediates and final products according to different transformation pathways. Among intermediates and final products, cellulose, lignin, biofuels, and simple sugars stand out as commodities, while some general examples of specialties are flavoring agents, intermediates for chemical synthesis, and building blocks for polymers. Most of these specialties come from further conversion of commodities via different conversion and separation routes. Similarly, to a refinery, these units start with a complex and multicomponent matrix (crude oil versus biomass) and its fractionation and conversion into a variety of products which will serve as feedstock to another industrial step. Different levels of conversion platforms can be considered in a single biorefinery, from high production (low price) to small production for a specific market (high price).

Pulp industry has been recognized as the leading industrial sector in biorefining since a long time ago due to the raw materials and to the integrated production of pulp (cellulose) and energy, mostly provided by burning lignin of the black liquor. In a more extended sense, pulp and paper industry is an example of realistic circular (bio)economy implementation, considering energy cycle, water recycle in the process, and chemical recovery cycle and on-site production. In the following sections, a general overview about chemical pulping will be given. The most common side stream and chemical recovery processes will be addressed along with the main steps for integration of biorefinery processes in pulp mills. Finally, recent advances of lignin end uses and new perspectives will be presented.

Keywords Pulp mills · Biorefineries · Lignin characterization · Delignification process · Radar tool · Kraft pulping · Sulphite pulping · Kraft lignin · Lignosulfonates · Organosolv · Bark

© Springer Nature Switzerland AG 2018
A. E. Rodrigues et al., *An Integrated Approach for Added-Value Products from Lignocellulosic Biorefineries*, https://doi.org/10.1007/978-3-319-99313-3_1

1

1.1 General Overview: Delignification Industrial Processes

Mechanical pulping and chemical pulping are the two main industrial processes for fiber separation, leading to materials with different properties for different end uses. Mechanical processes have a high energy demand, and fiber separation is achieved by mechanical force and softening the middle lamella lignin (Popa 2013). Chemical pulping is required for removing lignin from lignocellulosic matrix leading to the cellulosic fiber separation with minimum of mechanical force – delignification – and, therefore, this section is focused on these processes. This is the principle of pulping technology by chemical processes. However, nowadays, delignification has a broad sense, and the general concept has been applied to different types, rates, and extents of delignification in biorefining processes.

In pulping chemical processes, lignin is cleaved, and new charged groups are introduced leading to dissolution of lignin/lignin fragments in the pulping liquor, while the polysaccharide-rich fraction remains as a cake and proceeds for washing and bleaching stages. However, delignification has a limited selectivity for lignin; therefore, a fraction of initial carbohydrates, in particular the hemicellulose, is also dissolved in pulping liquor and lost. Pulping is stopped at a predefined level of residual lignin content in pulp in order to avoid further pulp yield decrease due to carbohydrate degradation. In general, chemical pulp fibers are more flexible than mechanical ones, leading to a better paper formation and giving good strength properties to chemical pulp. Kappa number is the parameter used for estimating the residual content of lignin in the pulp and the delignification degree.

Kraft Pulping
The first alkaline process developed for pulp production from wood is attributed to Burgess and Watt in 1851, known as the soda process. The introduction of sulfides into pulp production was patented in 1870 by Eaton, but the first kraft process was implemented only in 1885 in Sweden (Sjöström 1993). About 90% of the global chemical pulp is produced by kraft process. This process uses, as active reagents, sodium sulfide and sodium hydroxide. It is applicable to almost all wood species, resulting in high yields and pulps with better physico-mechanical properties compared to other processes and is therefore known as the "kraft process" ("kraft" stands for "strong" in German and Swedish). One of the disadvantages of this process is, besides the emission of malodor compounds, the dark color of pulp. However, the development of bleaching techniques in the 1940s has enabled the production of white cellulosic pulps of superior quality leading to the consolidation of kraft process. It is currently the dominant pulping chemical process in the world (90% of pulp production) (Popa 2013), being applicable to a wide variety of wood species with efficient regeneration of chemicals and energy (Clayton et al. 1983; Gullichsen 1999; Minor 1996; Ek et al. 2009; Sjostrom 1981).

The debarked wood logs are converted to woodchips and sieved, selecting the chips with the appropriate dimensions to assure uniform delignification at the digester. In the continuous process, the wood chips are pre-vaporized to facilitate impregnation and are introduced into the kraft digester with white liquor. White liquor, or pulping liquor, is composed of sodium hydroxide, NaOH, and sodium sulfide, Na_2S. The ions OH^- and hydrogen sulfide HS^- are the species that react with lignin during kraft pulping. The concentration of these chemicals is usually expressed in equivalent amounts of Na_2O, being effective alkali, active alkali, and sulfidity the most important parameters of white liquor. The active alkali concentration depends on the active alkali charge and also on the liquor-to-wood ratio. To produce bleachable pulp grades from hardwoods, the active alkali charge required ranges from 17% to 19% NaOH, while for softwood higher charge is required, usually between 20% and 25%. At higher values, for a fixed temperature, the delignification rate increases. In general, the initial active alkali concentration is 40–60 g/L NaOH for kraft pulping. White liquor includes also salts inactive in pulping process, such as sodium sulfate (Na_2SO_4) and sodium carbonate (Na_2CO_3), coming from chemical recovery cycle (referred later in this section). The sulfidity strongly influences the pulping rate and process selectivity, and it is expressed as the percent ratio between sodium sulfide and active alkali. The sulfidity ranges between 25% and 35% for the cooking of hardwoods and 35–40% for softwoods. Pulping temperature is in the range 160–175 °C and pressure 7–12 bar (Gullichsen 1999; Sjostrom 1981). Pulping is stopped by extracting the liquor containing the dissolved material, referred to as black liquor, and then washed, removing the uncooked chips and other fragments, obtaining the raw pulp. After that, impurities such as shives and dirt are removed in the screening process, and then the pulp can be bleached or used for the manufacture of paper or board. The black liquor and the washing water are then introduced in the recovery circuit: diluted black liquor is evaporated, and the concentrated stream is burnt at the recovery boiler for energy production from dissolved organic compounds and recovery of inorganic compounds. The resulting stream (smelt) results in green liquor which is converted to white liquor by causticizing, at the chemical circuit (Clayton et al. 1983; Gullichsen 1999; Ek et al. 2009). A simplified scheme of this sequence of processes is presented in Fig. 1.1. The pulp yield is approximately 50%, depending on the wood and conditions used (Gullichsen 1999). For each ton of pulp produced, about 10 tons of weak black liquor, containing between 1.2 tons and 2.0 of dry solids, are generated.

Kraft pulping operations are highly integrated, and the recovery process is one of the most relevant advantages assuring the sustainability of this process (Sect. 1.2). However, the recovery boiler capacity is limited, hampering the increase of pulp production. Therefore, some mills have been considering the removal of part of black liquor from the circuit, or organic material dissolved in black liquor, and, therefore, gaining some additional pulp production capacity. At the same time, several efforts have been made to get marketable kraft lignin product in order to cover the capital and operational expenditure, since burning is a very limited solution for lignin. Some companies have been designing and building their own lignin isolation processes, and there are commercial solutions for this purpose in the market, being already operational (Sect. 1.4.3).

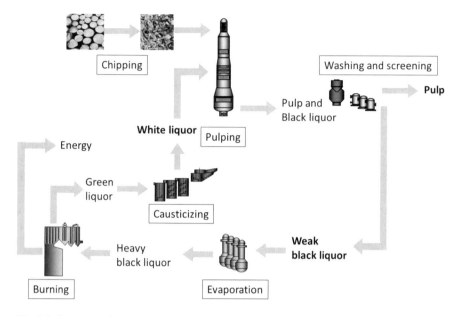

Fig. 1.1 Sequence of processes in kraft pulping

Sulfite Pulping

Sulfite pulping was developed and patented by the American chemist Benjamin Chew Tilghman in 1867. Acid sulfite pulping was the dominant pulping process until the 1940s, when it was succeeded by kraft cooking. Some modifications to the process were introduced, mainly in the 1950s and 1960s, replacing calcium by magnesium, sodium, or ammonium, giving rise to a more flexible process, extending it to different raw materials, and producing different pulps. The dominant cationic base in the process is magnesium. The sulfite pulping is referred to in different ways depending on the pH range used. In an acid medium, depending on the pH range, the process is named bisulfite acid (pH 1–2) or simple bisulfite (pH 3–5). In both cases, the active agents are H^+ and HSO_3^-. Neutral sulfite process operates at pH range of 6–9, and the active agents are HSO_3^- and SO_3^{2-}, while alkaline sulfite has a pH 10–13.5 with SO_3^{2-} and OH^-. The acid process produces dissolving pulp for textiles and pulp for newsprint or tissue, while pulps from neutral and alkaline sulfite are applied mainly for corrugated medium and package grades.

As compared with kraft pulp, cooking cycle is long, and it is not suitable for all wood species. However, acid sulfite pulp is brighter and can be used without bleaching to produce some printing paper grades. The main disadvantage of this process as compared with kraft one is the sulfite spent liquor due to the high sulfur dioxide loss and the unfeasibility of chemical recovery (Popa 2013). However, lignosulfonates have nowadays an established market being used mainly as dispersants in cement admixtures and dye solutions and binder for pelleting animal feed and for dust control, among others, thus creating an additional driving force for this process, besides pulp. Borregaard LignoTech is one of the world leaders of lignosulfonate-derived products, including vanillin flavor produced by oxidation of softwood lignosulfonates. Other player in the lignosulfonate market is Tembec, with an annual production capacity of 225,000 metric tons (Tembec 2002), recently acquired by Rayonier Advanced Materials (Rayonier Advanced Materials 2018). Burgo Group (Tolmezzo mill) is also a lignosulfonate producer, stating a wide range of applications (Burgo Group 2013).

1.2 Side Streams and Current Recovery Cycles of Chemicals and Energy in Typical Mills

Black liquor is the main side stream of kraft pulping, containing about one half of the wood content together with spent pulping chemicals. At the final stage of pulping, black liquor is separated from cooked woodchips, and pulp washing waters are collected together with main stream, giving rise to the weak black liquor. About 10 tons of weak black liquor is generated per ton of pulp produced, containing between 12% and 20% of total solids. Weak black liquor is evaporated to reduce the water amount to about 15%. The resultant stream, heavy (or concentrated) black liquor, is then injected into the recovery boiler, the most important component of the kraft recovery cycle. This process makes pulp production self-sustaining for energy and almost sufficient for pulping chemicals. Figure 1.2 summarizes the main steps of the kraft recovery process. Black liquor is one of the fifth most important fuels in the world: globally over 1.3 billion tons/year of weak black liquor is produced. This represents about 200 million tons of dry solids to recover 50 million tons of cooking chemicals and simultaneously producing about 700 million tons of high-pressure steam (Reeve 2002).

The recovery cycle is briefly described. At recovery boiler the organic content of the black liquor is burned producing heat, which is then recovered by heat exchangers and simultaneously takes place the first stage for chemical recovery. The heavy black liquor is injected; the organic material burns at 1000 °C, while other fraction burns onto the porous char bed under reducing conditions. In the char bed, sodium carbonate (Na_2CO_3) and sodium sulfide (Na_2S) are the main chemicals generated from the sodium and sulfur (mainly sodium sulfate, Na_2SO_4) coming with the black liquor. Sodium carbonate and sodium sulfide are the two major components of the smelt. Smelt (the mixture of molten salts) flows out of the boiler and is brought to the dissolving tank producing the green liquor. The recovery in modern mill is higher than 90%. Makeup sodium sulfate can be added to compensate losses in pulping, which in average are between 10 and 20 kg per ton of pulp. The green liquor is then clarified, removing insoluble components (dregs), and goes to the causticizing plant. At this stage, sodium carbonate reacts with quicklime (CaO) in the presence of water producing sodium hydroxide (NaOH) (conversion nearly 90%). Sodium hydroxide and sodium sulfide are the active components of the white liquor for the delignification reactions, and it is returned to the digester for reuse in

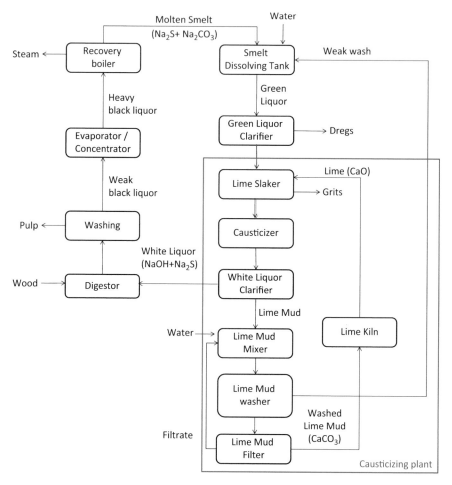

Fig. 1.2 Simplified flowsheet of the kraft pulping chemical recovery process

pulping. The CaCO₃ (slaked lime) resulting from causticizing is washed to reduce Na losses and sent to the lime kiln to regenerate CaO for reuse (Fig. 1.2).

Methanol is generated in the kraft cooking process. It comes from the delignification reactions, being washed out and then condensed at the evaporation plant and then often distilled in one or two distillation columns to produce a water-free methanol stream. The methanol stream contains various organic sulfur compounds and therefore has an unpleasant odor. Usually methanol is used as a support fuel for burners.

The bark and other forest biomass such as sawmill resulting from wood processing into fines and primary sludge and secondary sludge coming from effluent treatment are also considered side streams. Although not all of the forestry biomass coming from dedicated forest plantations is available for collecting (due to access and transport limitations and soil sustainability issues), there is an important source of lignocellulosic material that can be integrated into different value chains, generating products, materials, and bioenergy. The bark (produced in bole debarking process, being about 10% on wood basis) and forest biomass are currently burned at thermal boilers for energy production for using at mill site, and the surplus is introduced in the electrical grid. However, the potential of these solid side streams is well known as source of bioactive compounds, polysaccharides, and even lignin (Pinto et al. 2016). Primary and secondary sludge and sawdust are produced in lower quantities as compared with bark/forest residues. There are several opportunities for their applications such as fertilizers or as energy source as well. The strategy for their valorization is quite dependent on the particular company, being the landfill disposal the last choice for most of them for environmental and economic reasons.

1.3   The Integration of New Biorefining Processes in Pulp Industries

Forest-based industry is historically centered on pulp and energy production. Pulp and paper companies are mature commodity industries. However, several developments have been changing this paradigm, leading these companies to invest more in R&D activities and innovation and in a stepwise increase of business areas and product portfolio.

Biorefining and related processes have been considered and planned from a long time ago. Myerly and co-workers have published in 1981 the "Forest Refinery" (Myerly et al. 1981) giving their perspective based on up-to-date developments. In this paper, the authors have made a brief state of the art about biomass conversion to products, defending that "…we would get greater flexibility for its use if we were to refine biomass into cellulose, hemicellulose, and lignin cuts, preserving the integrity of each to the maximum degree possible. These pure intermediates could then be converted to highest value." From this perspective comes up that pulp industry is already a biorefinery converting forest material to bio-based products (pulp). Moreover, conventional kraft industry produces renewable energy from lignin and recovers pulping chemicals to be reintroduced in the process, while sulfite industry produces also lignosulfonates. Authors already mentioned other bio-based products already in the market at that time such as vanillin and turpentine, highlighting also the great potential to be exploited. Since then, massive scientific and technological progresses were made, and, nowadays, some pioneer industries have been extending their portfolio to other bio-based products, with several successful examples of biorefineries.

The pulping industry has a quite privileged position to gradually move into a forest biorefinery. This sector has the logistic, knowledge, and operational and industrial systems to be at the front head of the forest-based bioeconomy. Some companies have already moved to biorefinery activity. One of the world's most advanced forest-based biorefinery is Borregaard, producing lignin products, especially cellulose, vanillin, and bioethanol for several economic sectors, from cosmetics to construction. Other pulp and paper industries in Europe and in the world are moving to bioproducts and advanced biofuels. Some examples of this new trend are the Stora Enso, UPM, and Domsjö in Europe and Suzano and Fibria in South America. In this case, the recent acquisitions of the Canadian company Lignol and the new company, Fibria Innovations, are some examples of development or integration of technology in pulp and paper companies focused on bioproducts.

Figure 1.3 presents examples of biorefining pathways integrated in pulping industry. In this scenario, after chipping, wood goes to the pulp mill, and other forestry biomass comes into the pre-extraction and purification sector. This sector could be a unit dedicated to extract and fractionate bioactive compounds or hemicelluloses. The solid residue left after extraction is then introduced in the pretreatment/hydrolysis unit. At this stage, depending on the material and the previous type of process, a pretreatment (partial or complete delignification or steam explosion, among others, to liberate the fiber material) followed by hydrolysis could be considered. The final result should be sugar solutions and lignin. Lignin is also resulting from wood pulping process. Sugars and lignin can be further converted, in independent routes, to advanced fuels, chemical building blocks, polymers, or other products in the same site or in other value-added industries. Examples of products from sugars are ethanol, lactic acid, furfural, and hydroxymethylfurfural, while lignins can be converted to advanced fuels, phenolic resins, polymers, bio-based materials

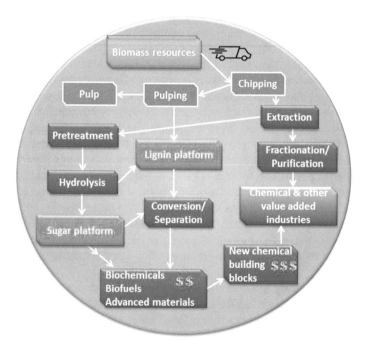

Fig. 1.3 Integration of new biorefining possible pathways in pulp industries

as composites or chemical intermediates, and flavors. In addition to these processes and products, other examples of future opportunity for pulp and paper mills are CO_2 capture, storage, and conversion (Jönsson and Berntsson 2010).

Considering the complex array of process and product opportunities, the evaluation of the biorefinery pathway must be investigated for each mill, taking into account the techno-economic risk assessment for factors such as biomass availability and price fluctuations due to competitive market for biomass, operating costs for new processes, and impact of technical solutions on product yield and quality and market. Defining the best business model looking to the best partner/product/process combination for optimization of profit margins is essential.

Biorefinery implementation and new investments are based on company's expectations in high revenues and competitive advantage. However, the integration of a biorefinery model in a pulp industry should be based on a stepwise approach to mitigate the associated risk. This concept was described for lignin-based biorefinery strategy (Téguia et al. 2017). According to the authors, this strategic implementation is divided into three phases of increasing financial benefits and risk:

Phase I – The short-term phase (<5 year) should provide technology and business risk mitigation; in this phase the core business does not change. The objective of biorefinery projects at this stage is to use the products in internal processes, aiming at reducing production costs or carbon foot-printing, for example, by producing fuel substitutes and providing early profits. The advantage of this stepwise implementation is the flexibility to move the orientation of phase II according to the market trends and in the case of a more valuable strategy is identified.

Phase II – The bio-based products will be manufactured and introduced in the market. This involves changes in the core business and mission and requires new capital, new technologies, and new product logistic, sells, and marketing. At this stage, a solid and strong biorefinery partner outside the forest/pulp sector is required. The objective is to increase revenues.

Phase III – This long-term strategy should be designed before embarking on phase I. At phase III the objective is to maximize the margins by scheduling product manufacturing by advanced supply chain techniques and requires manufacturing flexibility for conventional products (pulp, paper) and also for new bio-based products.

1.4 Lignin: The Main Side Stream from Delignification Process

1.4.1 Types of Lignins and Up-to-Date Market

The Confederation of European Pulp and Paper Industries (CEPI countries) has produced in 2016 about 41.7 million tons of pulp in the 153 pulp mills (CEPI 2017). The estimated value for lignin dissolved in black liquor per ton of bleached kraft pulp produced is between 400 and 500 kg. Considering CEPI's pulp production data, 17 to 20 million tons of kraft lignin dissolved in black liquor is annually produced in Europe. Globally, the production of lignin by kraft pulp mills is 78 million tons, most of them burned for steam and energy in the recovery process at pulp mill sites (Sect. 1.2). The current capacity for kraft lignin recovery is about 160,000 tons; however, in 2016, the effective lignin isolation was limited to 75,000 tons (Miller et al. 2016). The current trade-off between burning the black liquor (and lignin dissolved on it) and the recovery of some of the produced lignin is dictated by market prices and decisions on new investments. Pulp mills have been expanding their capacity by overcoming limitations, reaching to the last one requiring the highest investment, the recovery boiler. This is the so-called bottleneck of pulp mill increase capacity. In some cases, removing lignin from the recovery boiler load justifies lignin isolation and recovery, as far as the lignin product has an added value or, at least, the energy equivalent. The energy load is therefore removed, and isolated lignin can be stored and transferred for burning in another unit in the mill or sold for this purpose. On the other side, some companies have made investments in turbo generators to produce electricity, with eventual "overpower" being sold to the local grid. In fact, several pulp mills (in particular those that are not integrated with paper mill, which consume stream and energy) are self-sufficient and are producers of *green* power. These companies face the decision between selling lignin and selling electricity. The decision on new investment will be based on these three factors: (1) recovery boiler capacity and investment required to increase pulp production capacity, (2) current existence of turbo generator, and (3) investment required for lignin recovery unit, which in turn, will be dictated by energy price, new applications, and market for lignin products besides burning it to power production.

The production of lignosulfonates was, in 2015, about 1.1 M tons (potential available 3 M tons), and kraft lignin was 75,000 tons (potential 78 M tons), while soda lignin and biorefinery lignin account together for less than 10,000 tons, although a huge potential already exists, in particular for cellulosic sugars and ethanol biorefineries which are operating below capacity (Miller et al. 2016).

In 2014, Borregaard LignoTech (Europe), Tembec (Canada), and Aditya Birla Group (Asia, acquired the Swedish company Domsjö in 2011) dominated the global lignin market share, mainly attributed to lignosulfonates (sulfite process). In fact, only 2% of total lignin produced is commercialized. Lignosulfonate is still the main product, accounting for over 85% of global lignin market (1 M ton/year), but is rather targeted to specific market segment: the large markets are animal feed binders, additive for concrete admixtures, as well as dust controllers, while the medium and small markets are dyestuff and gypsum wallboard dispersants, resins, binders, and vanillin. For many decades MeadWestVaco, now Ingenity (WestRock Co.'s MWV Specialty Chemicals division), was the exclusive producer of kraft lignin and some derivatives such as sulfonated and aminated lignin for dispersants with application in textile dyes, agricultural chemicals, and asphalt emulsification. In the recent years, the capacity and production of kraft lignin have been increasing, reaching now to about 160,000 ton/year (Miller et al. 2016) – Table 1.1. Some companies produce sulfonated kraft lignin to increase water solubility and other properties for

Table 1.1 Lignosulfonate and kraft lignin capacity and market share reported for 2015 (Miller et al. 2016)

Company	Country	Capacity, thousand ton
Lignosulfonates		
Borregaard LignoTech	Norway	600
Domsjö Fabriker	Sweden	120
Nippon paper	Japan	100
Weili group	China	90
Wuhan East China chemical	China	80
Tembec	Canada	60
Shenyang Xingzhenghe chemical	China	50
Kondopoga, Vyborgskaya, Others	Russia	160
Kraft lignin		
Stora Enso Sunila	Finland	50
Ingevity	USA	50
Domtar	USA	25
Suzano	Brazil	20
West Fraser	Canada	10
Rise-bioeconomy	Sweden	Demo
Resolute	Canada	Demo
Liquid lignin	USA	Pilot

some markets. Lignosulfonate from sulfite pulp mills is highly sulfonated being suitable for concrete admixtures, while sulfonated kraft lignin can be produced with a controlled level of sulfonation, and usually higher purity can be achieved (Miller et al. 2016).

As more kraft pulp and paper industries are investing in new processes for lignin separation from black liquor, the challenge is to develop the lignin end use markets, being aware on new opportunities and investing in R&D on lignin products and in optimizing lignin properties for specific end uses. The forecast for the next years is that the global lignin market will reach to USD 913 million (2025) (Smolarski 2012). This would be possible by the integration of isolation technologies in pulp mills and the forthcoming of new lignin grades from biorefining activity. Nowadays, high-grade kraft lignin market has been limited by production capacity which, in turn, is limited by the market.

1.4.2 Lignins from New Incoming Delignification Processes

Chemical pulp mills operated as bio-based industries producing cellulosic fibers, recovering energy and chemicals, and some of them selling energy and even by-products to be integrated in other industries as raw materials. These industries use delignification processes adjusted to the final product pulp. However, in a broader sense, biorefineries are a facility arising since the 1990s which are focused on chemicals and fuels from renewable resources; these facilities use deconstruction and multistep processes to get bio-based products to replace the current ones produced from oil: the scope ranges from platform chemicals (p.ex. sugars) to high value-added chemicals. As described in Sect. 1.3, some pulp mills are nowadays dedicated to bio-based products other than pulp and energy. However, dedicated plants producing cellulosic ethanol from lignocellulosic materials, particularly non-wood materials, have been the most visible ones, and the estimated potential for lignin production from these second-generation cellulosic ethanol plants is about 250,000 tons (Miller et al. 2016). The first step in the process flow for sugar platform (precursor for ethanol) is based on organosolv and related processes or temperature and pressure action (steam explosion and autohydrolysis): these are the deconstruction or pretreatment processes. The next step is enzymatic or acid hydrolysis to get polysaccharide conversion to simple sugars. In general, deconstruction processes or a process to open the lignocellulosic matrix before hydrolysis (pretreatment) is required in order to increase the enzymatic digestibility. For all of these cases, a side stream of lignin is produced (Fig. 1.4).

If the pretreatment involves a delignification process (p.ex. organosolv), the lignin is produced at an early stage. If the pretreatment is a process to open the matrix and therefore the delignification is low at this first stage (p.ex. autohydrolysis and steam explosion), the lignin is recovered at the end of the process, after the hydrolysis. A third situation is possible, as the technologies already exist, as, for example, the CASE™ process, based on total hydrolysis of lignocellulosic material developed by Virdia (the former HCL CleanTech). After a pretreatment aiming to remove extractives and ashes, the complete wood hydrolysis is performed with concentrated hydrochloric acid (HCl). In 2014, Virdia was acquired by Stora Enso (Biofuels Digest 2014).

Another example is the Plantrose® process, a technology developed by Renmatix for cellulosic sugar production. The first step is called "hemihydrolysis"; it comprises a pretreatment based on autohydrolysis to solubilize and separate hemicellulose

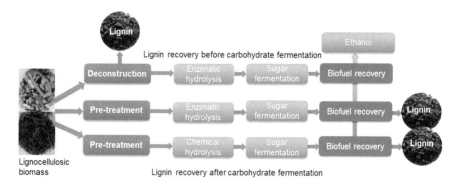

Fig. 1.4 Process flow approaches to produce sugars and ethanol produced based on enzymatic or chemical hydrolysis to convert biomass into sugars

and getting the resulting monomer (xylose in the case of hardwoods). Then the solid stream (containing cellulose and lignin) is submitted to supercritical hydrolysis using supercritical water acting as both a solvent and a catalyst, achieving cellulose hydrolysis into simple sugars. The remaining solids are composed of lignin that can be used for thermal value or for production of chemicals (Harlow 2016).

Delignification of wood and other lignocellulosic materials using organic solvents at high temperature has been extensively studied since the 1970s. Recently, new technological developments and the possibility of using this process as a pretreatment approach for biomass in biorefineries, operating at conditions to produce minimum lignin structure modification, brought new emphasis to organosolv process (Tao et al. 2016). The success of organosolv would be determined by the quality and quantity of products, by their application, and also by the recovery system for organic solvents.

One of the key advantages of organosolv pulping is the ability to produce organosolv lignin, a clean (low contaminant content) and less transformed lignin as compared with kraft and sulfite lignin. There are several organosolv pulping processes, using different solvents or mixtures and catalysts. The Alcell process™, based on mixture of ethanol/water by autohydrolysis, was the first one to operate in a demonstration plant. In this process, between 130 and 200 kg per ton of high purity lignin (0.5% sugar, less than 0.1% ash) is produced (Miller et al. 2016). The purity and some of the most important lignin properties and structural features have been published elsewhere (Costa et al. 2014). The Canadian company Lignol Innovation Corporation acquired the technology, and, more recently, in 2015, Lignol facilities and technology was integrated in Fibria Innovations, a subsidiary of Fibria Celulose SA in Brazil.

AVAPCO, an American process company, is also developing ethanol-based biorefinery technology (AVAP, American Value Added Pulping®) using ethanol and sulfur dioxide as pulping agents (AVAPCO 2011). At Thomaston Biorefinery, the company has a demonstration plant for AVAP technology and BioPlus® nanocellulose production, announcing recently new investments and partnerships (American Process Inc. 2015). In Europe, Chemopolis (Finland) (Chempolis 2017) provides technology based on organosolv process formico®, and CIMV (France) operates a plant in France and uses acetic acid/formic acid process, including a process to further delignify and produce cellulose (Snelders et al. 2014) (Fig. 1.5). The American Science and Technology (AST) claims that its proprietary organosolv process produces high-quality pulp, lignin, and also fermentable sugars (American Science and Technology 2017).

In addition to processes based on acetic and formic acid for organosolv fractionation, there are other variants, such as organosolv in alkaline medium (organocell, ethanol-alkali) or even oxidative (Milox, using peroxyformic acid) (Popa 2013), and an alkaline sulfite anthraquinone methanol pulping (ASAM process). A recent review about this subject was published elsewhere (Kumar and Sharma 2017).

Several publications have been reporting organosolv for annual plants or agricultural-derived biomass, such as wheat straw (Fig. 1.5), including recently new perspectives for lignin (Lange et al. 2016) and for pulp (Barbash et al. 2017).

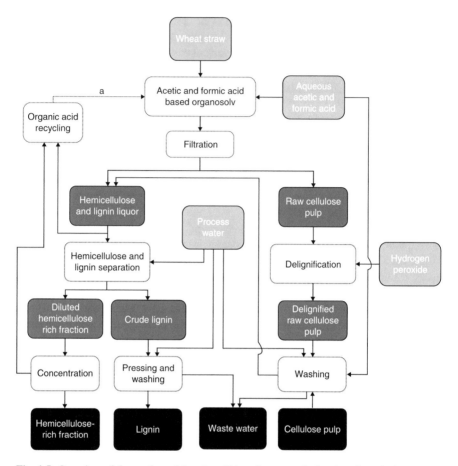

Fig. 1.5 Overview of the acetic and formic acid-based organosolv fractionation of wheat straw and its derived fractions. (Reprinted from Snelders et al. (2014), Copyright (2014), with permission from Elsevier)

However, organosolv process has still been investigated concerning solvent recovery (Silva et al. 2017), and the publications have demonstrated some progresses on conditions to bring together polysaccharide fraction recovery and end use and lignin quality (Panagiotopoulos et al. 2012). Figure 1.6 shows a flow diagram for a typical organosolv process, in this case, using ultrafiltration as downstream process for lignin aiming its fractionation (Alriols et al. 2010). A recent study about organosolv pulping as pretreatment process for producing ethanol and lignin from hardwood chips (Kautto 2017; Kautto et al. 2013) has reported higher energy demand due to the solvent recovery, as well as higher investment cost than the reference dilute acid process. The main conclusion of the economic assessment was that the lignin is the key factor to make the organosolv cost-competitive for this case. The final lignin applications will determine the price and, consequently, the

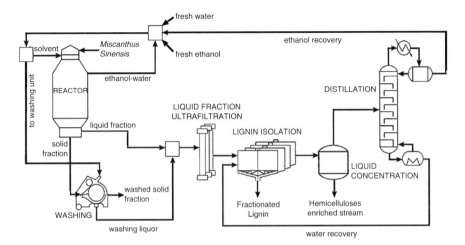

Fig. 1.6 Diagram of organosolv process, including a fractionation process for lignin by ultrafiltration, showing the flow of materials and solvent. (Reprinted from Alriols et al. (2010), Copyright (2010), with permission from Elsevier)

process viability when considering producing ethanol from polysaccharide fraction. However, the case can also be analyzed on the perspective of the polysaccharides, taking higher market value products than ethanol considering that the raw material is wood.

Biorefineries have been looking for new applications for lignin, since it represents 25–30% of the raw material left after conversion and could be an asset for the sustainability of these facilities. Biorefineries producing cellulosic ethanol have been facing some difficulties (Dale 2018). DowDuPont (USA) has recently announced its intention to sell the cellulosic biofuel business. The largest facility in the world to produce cellulosic ethanol opened in 2015, and the technology is based on hydrolysis based on diluted ammonia. However, other companies have been strengthening their activity. Examples are the POET-DSM Advanced Biofuels, LLC (USA), using a two-stage diluted acid hydrolysis (POET-DSM Advanced Biofuels 2014) and the Raízen Energia Participacões S/A (Brazil) operating a commercial biomass-to-ethanol facility using Iogen-modified acid-catalyzed steam explosion technology, close to the sugarcane mill (Iogen Corporation 2015). In Europe, Clariant owner of Sunliquid® technology is investing in a new full-scale commercial plant for the production of cellulosic ethanol from agricultural residues in Romania and setting up the Business Line Biofuels and Derivatives (Clariant 2018).

In spite of the great potential for lignin recovery as a valuable product, at present, most of the lignin generated in biomass biorefining processes is burned as fuel (Valdivia et al. 2016).

1.4.3   The Cost and the Revenues of Lignin Separation from Liquid Side Streams in a Pulp Mill

In a kraft pulp mill, the processes are highly integrated and dependent of black liquor (containing about 150 g/L of dry solids), which has a higher heating value (HHV) of 26 MJ/kg (dry solids). Black liquor is evaporated and burned to recover inorganic pulping chemicals and to produce steam and power from the organic components. In some cases, the recovery furnace is the bottleneck of the mill, limiting the pulp production. The solution could be an investment to upgrade the recovery furnace or, instead, divert black liquor or remove part of the dissolved lignin, reducing the thermal load. In this last perspective, some technologies for lignin isolation from kraft black liquor have been investigated and come on the market; simultaneously, recent developments in lignin-based products have been increasing industrial awareness on this polymer, and new perspectives have been created.

In the last decade, different research lines for lignin isolation from kraft liquor have given rise to two commercial processes: LignoBoost™ (Fig. 1.7) and LignoForce™ (Fig. 1.8). In the LignoBoost™ process (Axegård 2016), the lignin is precipitated by acidification with carbon dioxide and filtered through a filter press. The resulting cake is resuspended and acidified and filtered again and washed by displacement. Valmet has acquired this technology for development and commercialization. The first unit was installed at Domtar (Plymouth, North Carolina) in 2013 for a capacity of 25,000 tons of dry lignin (65% fry solids)/year. In an early stage, most of the lignin was used as fuel for energy, but now as much as 50% of its lignin production is for added-value applications (Miller et al. 2016) with UPM

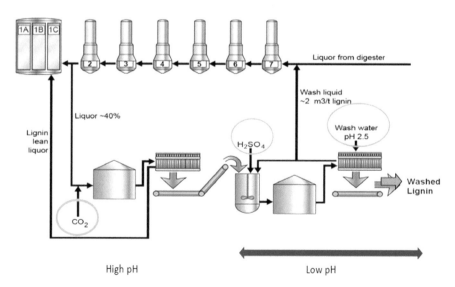

Fig. 1.7 Diagram of LignoBoost™ process owned by Valmet. (Reprinted from Axegård (2016), Copyright (2016))

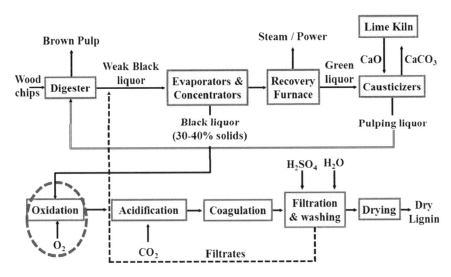

Fig. 1.8 Diagram of LignoForce System™ developed by FPInnovations and NORAM. (Reprinted with permission from Kouisni et al. (2016). Copyright (2016) American Chemical Society)

being the exclusive distributor of the commercial product, BioChoice™ in Europe (Domtar 2013).

In 2015, Stora Enso Sunila plant also installed LignoBoost™ technology with a production capacity of 50,000 tons/year of fir and pine lignin and lignin dust burners in the lime kilns; currently lignin is being marketed for phenolic resins (Stora Enso 2018), although other applications are now under evaluation (Stora Enso 2015). Valmet in collaboration with RISE Bioeconomy has developed a process for the reduction/elimination of lignin odors, making it suitable for other applications such as composites and bioplastics (Wallmo 2017).

The LignoForce System™ was developed by FPInnovations and Noram Engineering and Constructors Ltd. (Fig. 1.8) (Kouisni et al. 2016; Kouisni et al. 2012). This process is based on the acidification of black liquor, but there is a previous stage of oxidation with molecular oxygen under controlled conditions, and the second phase of acidification is performed in situ directly through the filter cake. The claimed advantages are (i) improvement of filterability of the acid-precipitated lignin; (ii) minimization of total reduced sulfur compounds, thereby leading to lower emissions of malodorous compounds; and (iii) the carbon dioxide and sulfuric acid requirements are significantly reduced because the oxidation of compounds and sugars consumes residual effective alkali (sodium hydroxide). There is a demonstration unit at Resolute Forest Products (Canada) for the development of processes with a production of 12.5 kg/h. The first commercial unit started operating in 2016 in West Fraser (Alberta, Canada) with a production capacity of 10,000 tons/year (Fraser 2018; PLANT 2016). Lignin Enterprises LLC (also known as Liquid Lignin) is an exclusive sales and marketing agent in the United States (Miller et al. 2016).

The Liquid Lignin Company (USA) presents another process for lignin isolation, the Sequential Liquid-Lignin Recovery and Purification, SLRP™ process (Fig. 1.9), patented in 2011 (Lake and Blackburn 2011; Lake and Blackburn 2014). The principle of this technology is to acidify with CO_2 at high temperature and pressure recovering a dense-phase "liquid lignin" from a black liquor stream. The washed-out black liquor stream returns to the evaporator system at a higher temperature than the black liquor taken to SLRP. The bulk liquid-lignin phase is then separated by gravity and acidified to pH 2–3 with H_2SO_4. The inventors claim energy saving as compared to other processes requiring dropping the black liquor temperature to precipitate filterable lignin and lower capital investment and operating cost (Gooding 2012; Lake and Blackburn 2014).

It is important to note that there have been some developments to associate condition adjustments in the isolation processes to meet technical requirements according to lignin applications. This is an ongoing research field for recognition of these technical requirements, which is the real challenge, considering the complex nature of lignin. In this sense, downstream research and development programs have been established for the commercial success of new lignin grades coming out, involving end users, producers, and research institutions with recognized expertise. This subject will be returned later in this subsection.

Kraft lignin can be modified for specific applications. The most common transformation is sulfonation or carboxylation for use as a dispersant. Other transformations are the partial removal of methoxyl groups or increasing the hydroxyl phenol

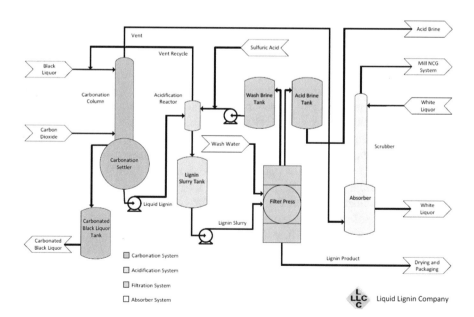

Fig. 1.9 Diagram of SLRP™ process (Sequential Liquid-Lignin Recovery and Purification) developed by Liquid Lignin Company. (Reprinted from Lake (2014), Copyright (2014), with permission from Romanian Academy Publishing House, the owner of the publishing rights)

content to improve its suitability for use in phenol formaldehyde (PF) resins, esterification for use in thermoplastics or coating applications, oxypropylation to use as bio-based polyol in polyurethane (PU) foams for insulation purposes, and partial depolymerization for performance improvement, for example, in PF resins and in PU foams.

Lignin price in the market depends on the lignin source, isolation process, and post-isolation treatment, as well as the final end use. Figure 1.10 shows the estimated lignin value and the respective potential product market value, differentiating between low purity, high purity, and specialty segment for lignin (according to the investment in purification/fractionation and lignin modification). Lignosulfonates and kraft lignin can be produced at different levels of purity and fractions, depending on the final end use. However, the efforts to produce high-grade lignin from a complex mixture are restricted by the end use, due to the associated high cost (CAPEX and OPEX). This is the reason for lignosulfonates and kraft lignin being at the lower level of lignin value, but there is some potential for upgrading and moving up higher segments (Fig. 1.10).

For conventional pulp mills that have been looking forward to increase the product portfolio and new biorefineries, the decision on investments for lignin isolation is driven by the lignin value and market forecast. Oil price is the most important driving force for the future success of lignin for both producers and end users:

1. Phenol-derived chemicals like polyols, resins, or even vanillin or derivatives ranging from medium to high price can be produced from oil or from lignin; as

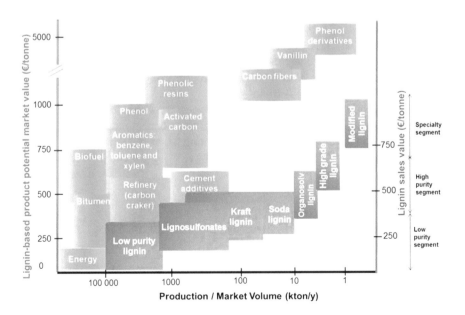

Fig. 1.10 Lignin value and corresponding lignin-based product, by lignin type and grade

the oil price rises, lignin becomes more attractive depending on the progresses on R&D and successful demo cases until this take place.

2. Pulp mill and biorefinery decisions between burning lignin for fuel (isolated or in black liquor), producing energy for internal use and for sale, or producing lignin for value-added application and purchasing fossil fuels, are mainly driven by oil price.

In all these cases, solutions and engineering studies for lignin conversion and demo cases should be a continuous process supported by government funds and some industrial clusters, particularly those that are looking for the short-term phase with low risk in biorefinery implementation (see Sect. 1.3). This has been the current practice, with increasing the number of demonstration projects and pilots around the world. Moreover, oil prices and environmental issues are concerns for biorefineries to operate to produce sugars, ethanol, and other commodities, being lignin a by-product whatever the final end use. However, industrials and researchers must be aware of lignin complexity and variety: kraft lignin derived from a certain wood species, isolated by a process in a pre-defined condition, could be suitable for an application, while a lignin coming from an alternative deconstruction process, resulting as a by-product after enzymatic hydrolysis, is different and not suitable to the same application. Even from the same source, the isolation conditions impart different properties to the final product (Costa et al. 2015).

In the last decade, there have been numerous studies of new applications for kraft lignin, recognizing the importance of the isolation process in the application. The isolation steps determine the contaminant content (components of the liquor that are coprecipitated with lignin), and also the lignin structure, which affects the reagent/ aptitude, required to develop a product. It has thus been emphasized the importance of aligning the purity/characteristics of lignin with the requirements for each application (Ahvazi et al. 2016; Costa et al. 2015) (Fig. 1.11). Some studies go further and also consider the fractionation of lignin in the sense of reducing the heterogeneity (Huang et al. 2017) or favoring a certain property, for example, hydroxyl group content or molecular weight distribution.

It is important to note that the isolation processes, with a fine tune of pH and other process parameters, could allow the lignin fractionation according to their properties due to their different phase partitioning behavior (Stoklosa et al. 2013).

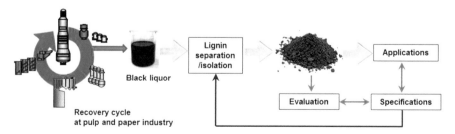

Fig. 1.11 Design of lignin value chain: black liquor, separation, characterization, and evaluation according to the product development specifications and improvement of separation process

By producing lignin fractions with different characteristics, the processes can be used to tailor the properties for different applications (Gosselink 2011). Moreover, it could also contribute to narrow the variability range of each fraction.

Recently other approaches for isolation or fractionation have emerged from scientific community. Some of the cases are revisiting processes already evaluated, such as the process of black liquor electrolysis (Oliveira et al. 2016) and membrane technologies (Costa et al. 2018; Humpert et al. 2016), now based on new knowledge and technological advances. Others are new and based on less conventional approaches, such as the case of direct extraction of black liquor with deep eutectic solvents, reporting good recovery rates under moderate conditions (Bagh et al. 2017). Other studies have been reporting the suitability of lignin for some applications according to their characteristics (Ahvazi et al. 2016; Kun and Pukánszky 2017). These few examples on progresses show that the scientific community is alert to new opportunities in this field.

1.5 Lignin Characterization and Classification

Lignin occurs widely in the middle lamellae and secondary cell walls of higher plants and plays a key role in constructive tissues as a building material, giving it its strength and rigidity and resistance to environmental stresses (Behling et al. 2016). This three-dimensional phenolic macromolecule is one of the principal components of the lignocellulosic materials (Fig. 1.12), together with cellulose and hemicellulose, and contributes as much as 30% of the weight and 40% of the energy content of lignocellulosic biomass (Azadi et al. 2013). The lignin content may vary in softwoods from 18% to 33% and in hardwoods from 15% to 35% (Azadi et al. 2013; Lin and Dence 1992; Ragauskas et al. 2014). In non-wood crops, such as cotton, cereal straws, flax, jute, bamboo, or bagasse, lignin content is generally lower and ranges from 5% to 30% (Buranov and Mazza 2008; Gellerstedt and Henriksson 2008; Monteil-Rivera et al. 2013).

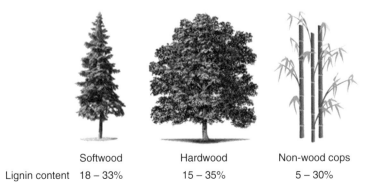

	Softwood	Hardwood	Non-wood cops
Lignin content	18 – 33%	15 – 35%	5 – 30%

Fig. 1.12 Lignin content in the major groups of lignocellulosic materials

In addition to the content also, the composition and structure (linkages and functional groups) of lignin may vary depending on the major groups of higher plants and also between species and even morphological parts of the same plant (Costa et al. 2016; Sevastyanova et al. 2014). Consequently, a full valorization and conversion of lignin into valuable products require a complete study about their characteristics and structure since the different kinds of linkages in this complex macromolecule and the diversity of their functional groups, such as methoxyl, phenolic, and aliphatic hydroxyl groups, have a great impact on its reactivity toward a further chemical process (Constant et al. 2016; Costa et al. 2014).

Lignin structure has been studied intensively for many years, although it is not yet fully understood. Besides all the existing information about its structure, lignin cannot be described through a simple structural characterization due to its high complexity. Lignin exhibits a complex three-dimensional amorphous structure, arising from the polymerization of its general structural subunit, the phenylpropane unit (ppu). The ppu can comprise several functional groups, being the most frequent ones aromatic methoxyl and phenolic hydroxyl, primary and secondary aliphatic hydroxyl, minor amounts of carbonyl groups (of aldehydes and ketones), and carboxyl groups (Calvo-Flores and Dobado 2010; Pinto et al. 2012). Coniferyl, sinapyl, and p-coumaryl alcohols are the precursors of the main moieties of lignin structure, guaiacyl (G), syringyl (S), and p-hydroxyphenyl (H), respectively, and differ between them in the methoxylation of the aromatic nuclei (Fig. 1.13).

The structural specifications of lignin have different implications on its reactivity, and for this reason, it is essential that for lignin characterization, several complementary methods have to be applied in order to determine their major structural features (Costa et al. 2015). Over the last decades, both destructive and nondestructive methods have been developed and applied for lignin characterization (Lin and Dence 1992). The chemical degradation methods include hydrogenolysis, nitrobenzene oxidation, cupric (II) oxidation, permanganate oxidation, ozonation, thioacidolysis, and also derivatization followed by reductive cleavage (Lu and Ralph 1997; Pepper et al. 1967; Quesada et al. 1999). These methods could provide information regarding the structure of lignin through the generation of low molecular weight compounds. However, all of them only comprise the selective cleavage of a specific fraction of lignin, hindering the study of the whole lignin structure. To overcome this challenge, various nondestructive methods are available and enable the identification and quantification of the main structural features of lignin (Lin and Dence 1992; Mansouri and Salvadó 2007). The advantage of nondestructive methods over degradation techniques is their ability to analyze the whole lignin structure and directly detect lignin moieties and/or functional groups. The nondestructive methods include different spectroscopic techniques such as Fourier transform infrared spectroscopy (FTIR), Raman spectroscopy, and nuclear magnetic resonance (NMR) (Barsberg et al. 2005; Capanema et al. 2004; Faix et al. 1994; Fernández-Costas et al. 2014; Froass et al. 1998). Among all, NMR spectroscopy is the most widely used method and provides detailed information about lignin molecular structure, both in terms of inter-unit linkages and functional groups. NMR techniques frequently applied comprise advanced one (^{13}C, ^{31}P and ^{1}H NMR)- and

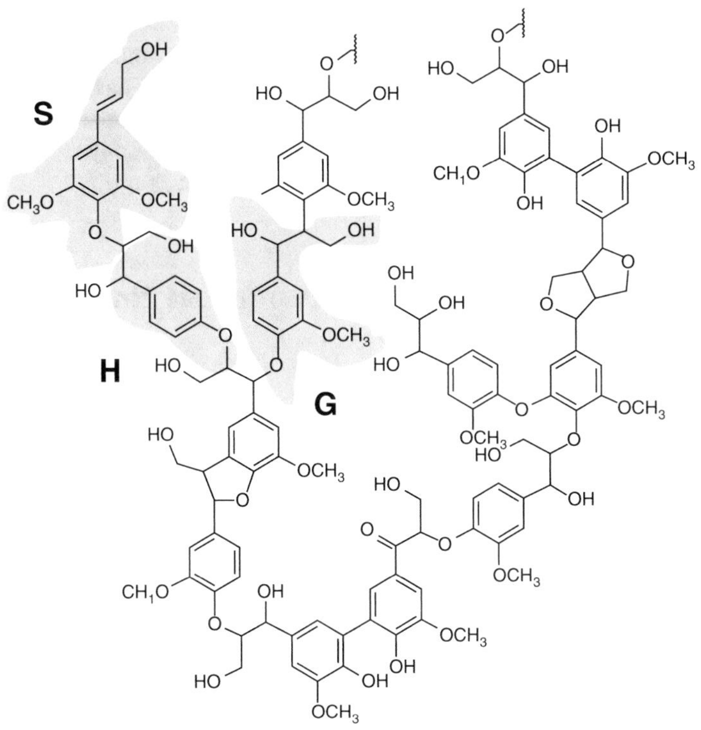

Fig. 1.13 Lignin structure with S, G, and H units

two-dimensional (2D) heteronuclear single quantum coherence (HSQC) (Capanema et al. 2004; Fernández-Costas et al. 2014; Wen et al. 2012). Quantitative ^{13}C NMR is frequently applied for the evaluation of the main structural features of lignins from hardwoods (Evtuguin et al. 2001; Fernández-Costas et al. 2014; Pinto et al. 2002b), softwoods (Capanema et al. 2004; Nimz et al. 1981), and annual plants (Costa et al. 2016; Sun et al. 2011; Xiao et al. 2001). This technique provides important information about carbons in different structural and chemical environments in lignin structure. ^{31}P NMR is also employed for functional group determination, since it allows to identify and quantify each type of hydroxyl groups in lignin structure (Argyropoulos et al. 2009; Costa et al. 2014; Heitner et al. 2010). All the referred techniques and methods generate extensive data about lignin structure: number of H, G, and S units, distribution of inter-unit linkages and functional groups, as well as the degree of condensation (DC) of this polymer. The DC is an important lignin characteristic often correlated (negatively) with lignin reactivity. Most commonly, the DC is related with the lignin moieties linked with C-C linkages with other lignin units via C_2 or C_6 positions in the aromatic ring of S units, C_2, C_5, or C_6 positions of the aromatic ring of G units, and in the case of H units, also C_3 position is available (Berlin and Balakshin 2014). Despite the several analytical techniques available, the aromatic composition and inter-unit linkages in different lignins are still not unequivocally established, and the assessment and correlation between the lignin structural features are essential for the evaluation of its potential as source of value-added compounds (Constant et al. 2016). Consequently, a multi-technique approach is required to obtain essential knowledge on the lignin structure, composition (acid soluble and insoluble lignin, residual carbohydrates, and inorganic fraction), molar mass distribution, and thermal behavior.

1.5.1   Impact of Delignification Process on the Structure of the Lignin

In addition to the source of lignocellulosic biomass, also the delignification process has considerable influence on lignin structure (distribution of linkages, structures, and/or functional groups) and consequently on its reactivity (Pinto et al. 2011). Technical lignins, obtained as a result of lignocellulosic biomass processing, differ dramatically from the corresponding native ones as a result of a combination of multiple reactions that promote significant physical and chemical modifications even within the same species. Consequently, a detailed understanding of technical lignin composition and structural features is particularly important in order to direct the efforts toward their valorization.

Technical lignins can be classified from different points of view; the presence or absence of sulfur in lignins is one structural characteristic that can distinguish them. The main sulfur-containing lignins comprise lignosulfonates and kraft lignins, while sulfur-free lignins include soda, organosolv, and steam explosion ones. Sulfur-free lignins can be divided into two categories, lignins from solvent pulping (organosolv lignins) and from alkaline pulping (soda lignins), and all of them show valuable properties that make these lignins an attractive source of low-molar mass phenol or aromatic compounds (Laurichesse and Avérous 2014). Among various delignification processes, there are four dominant chemical pulping processes, namely, kraft, sulfite, soda, and organosolv (Table 1.2). Kraft and sulfite pulping process accounts for more than 90% of the chemical pulp production (Ahvazi et al. 2016).

The main reactions comprised in structural modification of technical lignins comparatively to native ones involve predominantly degradation and condensation reactions. Lignin degradation occurs predominantly via α-aryl ether and β-aryl ether linkage cleavage. These types of linkages are the most easily cleaved and lead to an increase of phenolic hydroxyl content, although the cleavage mechanisms are different for the different delignification processes. Moreover, lignin degradation also could lead to a decrease in aliphatic hydroxyls, oxygenated aliphatic moieties, and the formation of carboxyl groups and saturated aliphatic structures (Berlin and Balakshin 2014). In opposition to lignin degradation, some repolymerization and/or condensation reactions could also occur and modify lignin chemical structure, leading to an increase of lignin molecular mass and content of condensed structures and decrease of its reactivity (Berlin and Balakshin 2014; Pinto et al. 2012). In literature it is possible to find several studies about structural characteristics of lignins

Table 1.2 Classification and some characteristics and properties of technical lignins. (Berlin and Balakshin 2014; Laurichesse and Avérous 2014; Pinto et al. 2012)

	Kraft	Lignosulfonate	Organosolv	Soda
Raw materials	Hardwood and softwood	Hardwood and softwood	Hardwood, softwood, and non-wood crops	Non-wood crops, hardwood, and softwood
Scale of production	Industrial	Industrial	Pilot/demo	Industrial
Sulphur content	Moderate	High	Free	Free
Solubility	Alkali	Water	Organic solvents	Alkali
Purity	Moderate	Low	High	Moderate-low
Pretreatment chemistry	Alkaline	Acid	Acid	Alkaline

obtained from the main delignification processes, either from softwoods (Froass et al. 1998; Pan et al. 2005; Saito et al. 2014; Sannigrahi et al. 2009), hardwoods (Ibarra et al. 2007; Pan et al. 2006; Pinto et al. 2005a; Pinto et al. 2002a; Wen et al. 2013b), or both (Capanema et al. 2001), and also herbaceous plants (Alriols et al. 2010; El Hage et al. 2010). In general, the authors had find that lignins obtained from delignification processes also show higher contents of phenolic hydroxyl groups, carboxylic acid groups, and condensed structures and lower contents of aliphatic hydroxyl groups and β-aryl ether linkages than the respective native lignins (Fernández-Costas et al. 2014; Ibarra et al. 2007; Lourenço et al. 2012).

However, it is important to note that the type and extension of the discussed structural alterations are strongly dependent on the process conditions (temperature, solvents, time, pH, and others) and the feedstock origin of lignin (Berlin and Balakshin 2014). Thus, the whole knowledge about the lignin structural changes imparted by delignification process is essential for the improvement of lignin valorization as raw material for the production of value-added compounds.

Kraft Lignin

As already stated in Sect. 1.1, nowadays kraft pulping accounts for 90% of the world's chemical pulp production, and 50–55 million metric tons of lignin are produced annually in the form of black liquor (Mahmood et al. 2016).

Kraft lignins have characteristic structural features that distinguish them from native lignin and also from other types of technical lignins. The extension of lignin structural modifications and also composition depends fundamentally on kraft pulping conditions applied and biomass species studied. Several authors study the lignin depolymerization mechanisms that prevail during kraft pulping and have shown that ether bonds, such as α- and β-aryl, are primarily and extensively cleaved (Chakar and Ragauskas 2004; Costa et al. 2014). Consequently, the cleavage of these inter-unit linkages and also some biphenyl units leads to a larger amount of phenolic hydroxyl groups in this type of lignins (Costa et al. 2014; Sevastyanova et al. 2014). The cleavage of some lignin structures that occur during kraft delignification leads to lignins with lower molecular weight than native lignins; moreover, softwood kraft lignins' molecular weight is, in general, higher than hardwood kraft lignins

(Pinto et al. 2012). Other structural characteristic of kraft lignins is the accumulation of carbon-carbon-resistant linkages (highly condensed G units and some linkages related to condensed structures like β-5, 5–5', 4-O-5) as result of the severe cooking conditions applied comparatively to the respective native ones (Costa et al. 2014; Gordobil et al. 2016; Pinto et al. 2012). Kraft lignins usually contain 1–2% of sulfur by weight as aliphatic thiol groups (Ahvazi et al. 2016).

In literature, some works studied the structural characterization of several technical lignins and found that kraft lignin appears to be the best of all the studied raw materials for the production of wood adhesives due to the highest content of phenolic and aliphatic hydroxyl (Lora 2008; Mansouri and Salvadó 2006). Moreover, the authors also found that kraft lignin is the most reactive material toward modification, and consequently this lignin could be a good raw material for lignin derivatives such as chelating resin adhesives and especially phenol-formaldehyde resins (Lora 2008; Mansouri and Salvadó 2006). The main limitation for the use of kraft lignin in value-added applications is its poor quality with respect to its nonhomogeneous molecular weight distribution, the impurities originating from both the wood itself and process elements, and the varying amount of functional groups, resulting in reactivity differences.

Lignosulfonates

As already referred, kraft process has become the dominant pulping process in the world, and consequently the importance of sulfite pulping has drastically decreased (Azadi et al. 2013).

Sulfite pulping is based on the use of aqueous sulfur dioxide as well as calcium, magnesium, or sodium as the counterion. This process operates at low values of pH, and the resulting lignin (lignin sulfonate or lignosulfonate) has about 5% of sulfur as sulfonate groups (Ahvazi et al. 2016). The predominant reaction in sulfite pulping is sulfonation of lignin side chain, predominantly through the introduction of sulfonate groups in the C_α-positions via the intermediate formation of carbocations which leads to lignin hydrolysis (Azadi et al. 2013; Gellerstedt and Henriksson 2008). The degree of degradation, the molecular weight, and the number of phenolic hydroxyl groups that can be observed in lignosulfonates are strongly dependent on the reaction conditions applied during the sulfite process. However, lignosulfonates are typically highly cross-linked lignins; consequently their weight-average molecular weight is higher than kraft lignins, and values in the range 10–60 kDa have been reported in literature, with a corresponding broad polydispersity (Azadi et al. 2013; Berlin and Balakshin 2014; Laurichesse and Avérous 2014).

Considering lignosulfonate properties and structural characteristics, these lignins represent valuable raw material for application in several industrial applications such as binders, dispersing gent, surfactant, adhesives, and cement additives (Berlin and Balakshin 2014; Laurichesse and Avérous 2014).

Organosolv Lignin

As an effective delignification technique, organosolv process has been successfully applied to different types of lignocellulosic biomass: hardwood, softwood, and nonwood plants. In this process a mixture of organic solvents with water is used as pulping media at relatively low pH and elevated temperature. The most commonly used solvents for organosolv delignification include alcohols such as methanol and ethanol, organic acids such as formic and acetic acid, and mixed organic solvent-inorganic alkali chemicals (McDonough 1992). The most investigated organosolv process is the Alcell process deployed at industrial scale in Lignol (Eastern Canada) during the 1980s. This process can be carried out in aqueous ethanol liquor at moderated acidity (Berlin and Balakshin 2014; Laurichesse and Avérous 2014). Organosolv process produces high-quality lignins of low molecular weight (Azadi et al. 2013). Some studies in literature demonstrate that this process degraded the lignin to a noticeable extent; however data from ^{13}C NMR and FTIR spectra showed that the degradation reactions that occur during the process did not significantly change the core of the lignin structure, remaining considerably unaltered (El Hage et al. 2009; Mansouri and Salvadó 2006). Other studies about organosolv delignification showed that the resulting lignins contain considerable amounts of phenolic and aliphatic hydroxyl groups as well as condensed units (Ahvazi et al. 2016; Costa et al. 2015). Moreover, all of them stated that the cleavage of β-O-4 linkages was the major mechanism of lignin breakdown (Fig. 1.14), since these structures are more susceptible to hydrolysis during organosolv treatment (El Hage et al. 2010; Hallac et al. 2010; McDonough 1992). Wen and co-workers found, by ^{13}C NMR, that the content of $S_{3,5}$ and $G_{3,4}$ in etherified lignin decreases, while the amount of $S_{3,5}$ and $G_{3,4}$ in non-etherified lignin is increased after organosolv treatment, indicative of β-O-4 linkage cleavage (Wen et al. 2013a). In contrast, other inter-linkages, such as β-5 and β-β linkages, increase following the organosolv process, which indicates that some condensation reactions occur during this delignification process (Wen et al. 2013a). Organosolv lignins also are practically sulfur-free and relatively pure, with low amounts of inorganic contaminants, such as carbohydrates and ashes. The higher purity of the isolated lignins is a result of the conditions applied during organosolv process that cause an intensive hydrolysis of lignin-carbohydrate complex (Wen et al. 2013a; Xu et al. 2006).

The structural modifications considering molecular weight, chemical structure, and functional group distribution of the generated lignin derivatives depend on the organosolv process conditions, and despite the resulted characteristics, this type of technical lignins is one of the most suitable for further conversion and valorization. However, in spite of the referred advantages, so far, none of the organosolv pulping technologies has been yet widely adopted in a production-scale mill.

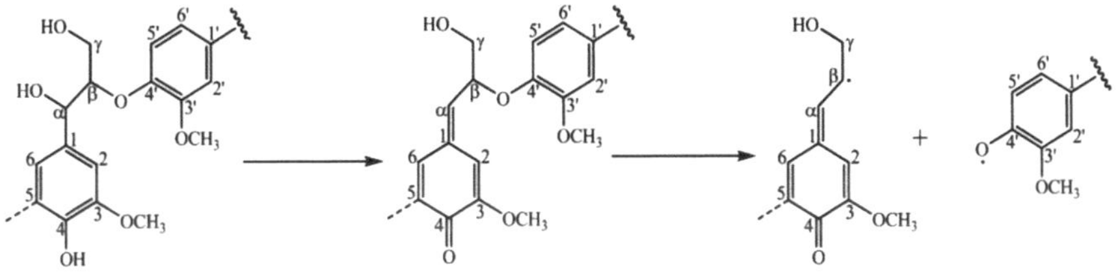

Fig. 1.14 Cleavage mechanism of β-O-4 linkages during ethanol organosolv process. (Reprinted with permission from Hallac et al. (2010). Copyright (2010) American Chemical Society)

Soda Lignins

Soda pulping process accounts for nearly 5% of the total pulp production; this process is mainly applied to non-wood plants such as straw, bagasse, kenaf, hemp, and, to some extent, hardwoods (Azadi et al. 2013; Lora 2008). Due to the lower content of lignin in non-wood plants, this pulping process only requires 10–15% of NaOH based on the raw material for delignification. In soda pulping, lignin depolymerization takes place mostly in non-phenolic β-aryl ether units (Fig. 1.15). The resulting lignin from soda process is often enriched with carboxylic acids and condensed structures. The former is due to the oxidation of aliphatic hydroxyl units, whereas the latter is caused by the absence of sulfide ions allowing the intermediate quinone methide structures to lose formaldehyde, causing the formation of undesirable alkaline stable enol ethers (Lora 2008).

Due to the absence of sulfur in their structure, the molecular weight of C_9 monomers of soda lignin is typically lower than that of kraft and sulfite lignins (Mansouri and Salvadó 2006).

As in soda pulping process, lignin extraction is based on hydrolytic cleavage of the native lignin; it results in a relatively chemically unmodified lignin comparative to other technical lignins (Laurichesse and Avérous 2014). Soda lignins are low molecular weight lignins and insoluble in water and contain small amounts of carbohydrates and ash contaminants resulting in a low to moderate lignin purity (Lora 2008).

1.5.2   Radar Tool for Lignin Classification on the Perspective of Its Valorization

One of the challenges associated with exploiting lignin structure and composition is the variability resulting from the type of plant and species, the delignification process, and the subsequent processing (Costa et al. 2015). All of which modifies its structure, making a constant and uniform lignin difficult to obtain. As already stated,

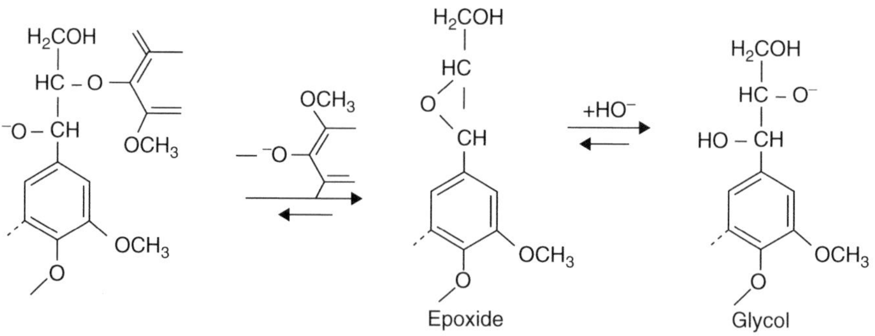

Fig. 1.15 Main reactions leading to the formation of soda lignins. (Reprinted from Lora (2008), Copyright (2008), with permission from Elsevier)

lignin complexity makes that data obtained from its characterization could not be easy to understand, making it difficult to extract focused and "ready-to-use" information. However, lignin structure and composition need to be comprehensively analyzed in order to provide a predictive tool of lignin's chemical and physical properties (Ragauskas et al. 2014).

Radar plots (Fig. 1.16) could represent an effective classification technique for lignins and are one useful approach for assessing their characteristics with the aim of maximizing lignin valorization. This tool has an immediate interpretation of the results and is easy to build, being frequently used for graphical representation of multivariate data in health, environmental, and pharmaceutical research (Picardo et al. 2013; Ritchie et al. 2011; Teixeira et al. 2010). In the case of lignin, radar classification can be adapted to include different characteristics according to the application required, being a useful predictive tool for product and process design in view of a particular valorization pathway. The selected lignin characteristics are the descriptors used to build the radar plot, and consequently, radar plot interpretation is strongly dependent on the selected descriptors and its displayed order. Moreover, it is assumed an equal contribution of each radar plot descriptor for the required objective.

Rodrigues and co-workers developed a line of investigation where radar plots were built to study vanillin and syringaldehyde production by oxidation of lignin in alkaline medium (Costa et al. 2015). Using selected descriptors identified as key characteristics, the authors evaluate the availability of different lignins for the production of these value-added phenolics. In several publications of the group (Costa et al. 2016; Costa et al. 2015; Pinto et al. 2016), lignins obtained from stalks and roots of herbaceous plants (corn, cotton, sugarcane, and tobacco) and lignins from different hardwoods (*Eucalyptus globulus*, *Acacia dealbata*, and *Salix* spp.) and from different parts of the same species (bole, bark, and branches of *E. globulus*), produced by different delignification processes (kraft and organosolv), provide the information sources used to build the lignin radar plots. To build the radar plot for each lignin, the authors selected as key characteristics the content on β-O-4 structures, noncondensed structures, S and G units, and the yield of Sy and V obtained by nitrobenzene oxidation. The selected descriptors reduce the unavoidable complexity of lignin structure to its key aspects while maintaining the scientific basis of the data sets with quantitative information (Costa et al. 2015). Moreover, the radar classification of lignins can be adapted to include different or additional descriptors being a useful predictive tool for product and process design.

In a recent work, the authors proceed to the experimental validation of the radar classification using two organosolv lignins from tobacco stalks: (1) butanol organosolv lignin and (2) ethanol organosolv lignin. The radar plots for tobacco lignin evaluation as source of phenolic compounds were built using the parameters β-O-4 and noncondensed structures, S and G proportion (drawn from S/G/H as total), and nitrobenzene yield on V and Sy (as a measurement of the reactivity of noncondensed fractions of lignin); the selected descriptors allow the qualitative prediction of the yield expected by oxidative depolymerization under the same range of conditions. Batch oxidation in alkaline medium with O_2 of tobacco organosolv lignins

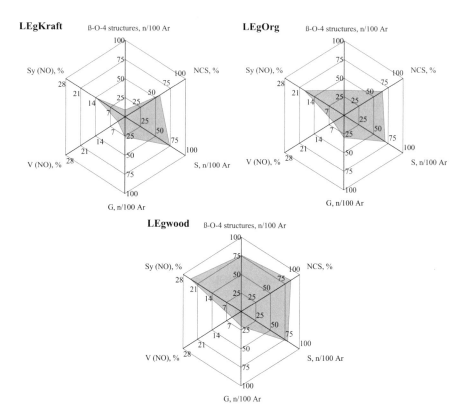

Fig. 1.16 Radar classification for eucalyptus wood lignins produced by different processes (NCS, noncondensed structures). Reprinted with permission from Costa et al. (2015). Copyright (2015) American Chemical Society

was performed and the products profile for each lignin, disclosing the maximum yields obtained for V and Sy (Costa et al. 2015).

Ethanol organosolv lignin produces 1.2% of V and 0.94% of Sy, while for butanol organosolv lignin, lower maximum values were obtained (0.74% of V and 0.34% of Sy), which is in accordance with the prediction provided by the radar classification using the selected descriptors (Fig. 1.17). Based on the radar classification, the ascending order of lignins according to the prospective yield for V and Sy by oxidation with O_2 in alkaline medium under the same conditions (pH, temperature, O_2 partial pressure) was butanol organosolv lignin < ethanol organosolv lignin. Considering the results, it can be concluded that oxidation with O_2 in alkaline medium of lignins from ethanol and butanol organosolv process under the same conditions has confirmed the qualitative differences of yields predicted by the radar classification (Costa et al. 2015).

The radar classification technique presented by the authors allows the screening of lignins resulting from industrial or preindustrial processes for their potential as source of phenolic compounds. Radar plots developed from key descriptors may

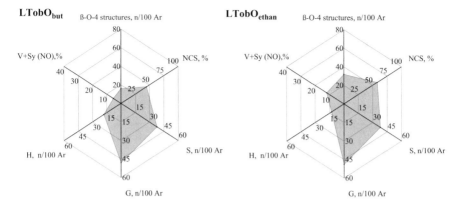

Fig. 1.17 Radar classification for tobacco lignins produced by organosolv process using butanol (LTobO$_{but}$) and ethanol (LTobO$_{ethan}$) (NCS, noncondensed structures). Reprinted with permission from Costa et al. (2015). Copyright (2015) American Chemical Society

demystify the complexity of lignin and direct process variable selection to achieve maximum valorization of lignin. This is one aspect to which lignocellulosic processors must pay attention, today and in the future.

1.5.3 Improving and Recognizing the Lignin Quality in Biorefineries

The second half of the twentieth century was dominated by the use of crude oil as raw material for energy, transportation, and chemicals. Technologies for processing crude oil have been developing since the 1860s, and refineries are highly integrated in industrial plants (Strassberger et al. 2014). However, within the past decade, motivated by environmental concerns and high oil prices, there has been an increasing interest in the concept of biorefineries focused in lignocellulosic biomass; consequently much of the biomass that is currently considered waste will become a valuable feedstock (Miller et al. 2016). The European Environment Agency estimates that Europe's biomass production capacity could grow up to 300 Mtons by 2030 (Strassberger et al. 2014). However, one of the biggest obstacles that holds back the development of biomass-based biorefineries is the efficient breakdown and conversion of lignocellulosic material to bio-based energy and high added-value chemicals (Laurichesse and Avérous 2014).

Lignin plays a significant role in the operational improvement of the emerging lignocellulosic-based biorefinery activity; it is available at large scale from the side streams of pulp and paper industries, representing a valuable renewable resource (Berlin and Balakshin 2014; Pinto et al. 2011; Strassberger et al. 2014). Annually, more than 75 Mtons of lignin are produced worldwide as by-product from wood pulping (Miller et al. 2016); as already stated the most abundant industrial lignins are the result of kraft and sulfite pulping processes. However, nowadays, most of the produced lignin is used as fuel for heat generation in the recovery boilers in pulp and paper mills. Whether lignin is burned for steam and energy or recovered as a chemical, there is an environmental benefit. Moreover, its fuel value has been estimated at $300/ton, while prices for kraft lignin have been estimated at $300–500/ton and lignosulfonate at $250–400/ton (Miller et al. 2016). The value of lignin is also dependent on the application; higher value applications can potentially bring higher prices for lignin since the lignin has further quality. For some applications, such as carbon fiber or fine chemicals, purer forms of lignin or chemically modified lignins will need to be developed, and these will be even more expensive. Consequently, the development of lignin market implies that biorefineries begun to improve different applications (Miller et al. 2016). In literature, the data point out that even if only 25% of lignin will be isolated from 600 Mtons of biomass (only half of the US prediction by 2060), it would mean that 150 Mtons of isolated lignin would be available in the United States (Strassberger et al. 2014). Without new product streams, the lignin introduced would far exceed the current world market for lignin used in speciality products.

1.6 Bark: An Unrecognized Valuable Lignocellulosic Material

Presently, chemicals and fuels are mainly obtained from fossil fuel, which is a resource that is becoming scarce, with frequent price oscillations, and encompasses serious environmental problems such as greenhouse gas emissions.

Researchers and industries are joining efforts to develop sustainable and environmentally friendly processes to overcome the strict demands for fuel, chemicals, and energy. This is in strong agreement with the emerging trend and concept of biorefinery of providing a better exploitation of the available renewable resources allowing the companies to increase their portfolio of products, obtain more profits, and be more environmentally friendly.

In the particular case of bark, it is a resource plentifully available in nature and is a byproduct generated in large amounts by pulp and paper industries since its utilization to produce pulp turns the pulping process economically unfeasible due to the low-quality pulps produced and high consumption of cooking chemicals (Miranda et al. 2012).

Currently, bark is burnt at the mill site to suppress power needs; however, in a perspective of integrated biorefinery applied in pulp and paper industries, several additional end uses to bark can be considered to increase the chain value of this industrial sector ranging from high-value low-volume products such as bioactive chemicals from low-value-high-volume products (a wide diversity of materials) (Feng et al. 2013).

The market for bark is in increasing expansion, and, in this regard, research in the last years has been devoted in understanding the complexity of physical and chemical composition of *E. globulus* bark in order to take out the most of this valuable resource (Cadahía et al. 1997; Kim et al. 2001; Miranda et al. 2013; Mota et al. 2013; Santos et al. 2012; Vázquez et al. 2009). Herein, the chemical composition of *E. globulus* bark will be detailed, and a brief discussion about possible valorization routes for bark is highlighted.

1.6.1 Chemical Composition: The Particular Case of Eucalyptus globulus *Bark*

Bark is a natural protector of the tree against physical and biological aggressions and corresponds to the external and surrounding tissues of the vascular cambium. It can be distinguished as the inner and outer bark: the inner bark is known for being the active layer and corresponds to the one nearest to wood and is mainly composed by secondary phloem. The external bark corresponds to the dead tissue containing the periderm attached to the outermost surface being mainly composed of old phloem and periderm (Fengel and Wegener 1984; Whitmore 1962). *E. globulus* bark anatomy has been extensively studied (Quilhó and Pereira 2001; Quilhó et al. 1999; Quilhó et al. 2000), and it is a dark-gray color bark, fibrous, deciduous, and with smooth, longitudinal, and narrow fissures; the parenchyma and fibers occupy 50% and 28% of total area/volume of bark, respectively (Quilhó et al. 2000). The inner bark region is composed of non-collapsed phloem, followed by collapsed phloem and a single periderm. The non-collapsed phloem is uniform and with alternate tangential bonds of parenchyma and fibers.

Considering the *E. globulus*, it is one of the most important species used in pulp and paper production in southern Europe, Australia, and South America. During the debarking process, about 11% of the stem dry weigh (Quilhó and Pereira 2001) is generated turning this particular bark a very promising biomass source, and therefore, several studies have been devoted to this particular species to understand its chemical composition and identify the main potential end uses for this plentifully available resource.

Chemical constituents of the bark can be grouped into structural components of the cell walls with several macromolecular components (polysaccharides, lignin, and suberin) and nonstructural components of low molecular weight (extractives and ashes) (Pereira 1988).

The chemical composition of *E. globulus* bark is summarized in Table 1.3. It is important to take into consideration that there is an inherent variability of the composition of barks related to geographic location, age, type of soil, and morphologic region studied, among other factors, explaining the wide range of intervals found between different sources of literature (Pereira 1988; Quilhó et al. 1999). Additionally, the extractive content can also vary according to methodologies applied and morphological part of the bark used (root, stem, or branch). The inner and outer bark have different characteristics, and thus, when comparing the chemical composition from different literature sources, it is important to take into consideration the morphological part of the bark studied (inner, outer, or both parts of the bark). The bark particle sizes also influence the extraction yields since mass transfer is improved. Some studies have shown that particle size influences the composition as well and that the finest fractions (<0.180 mm) are richer in extractives and lignin, while the coarser fractions (>2 mm) have higher cellulose and hemicellulose contents (Miranda et al. 2013).

1.6.1.1 Structural Components

Lignin

Lignin confers rigidity and resistance to the cell plant wall (Fengel and Wegener 1984) and is the second most representative fraction of bark components representing about 18.6–34.1% of the *E. globulus* bark dried weight (Mirra 2011; Mota et al. 2013; Pereira 1988; Vázquez et al. 2008) and where 16.7–26.6% is insoluble in mineral acids and only a small fraction (1.4–7.5%) is solubilized during acid hydrolysis.

E. globulus bark has a high ratio of S/G units, and, unlike its corresponding wood where H units are insignificant, about 7% of H units were detected (Costa et al. 2014; Mota et al. 2013). Deep characterization of *E. globulus* bark and wood lignins revealed that they are very similar, yet bark lignin contains higher content of condensed structures per aromatic ring (Costa et al. 2014). Costa et al. (2014) has proposed the following empirical formula for *E. globulus* bark lignin: $C_9H_{9.1}O_{2.3}(OH_{ph})_{0.31}(OCH_3)_{1.53}$, with phenyl propane unit molecular weight of 207 g/mol and OH_{ph} representing the phenolic groups.

Polysaccharides

Lignin and polysaccharides are the main components of the cell walls conferring mechanical strength, rigidity, and resistance to the plant. The polysaccharides are linked with lignin, and therefore, the type and proportion of sugars obtained might vary according to the depolymerization method employed (Chen 2014).

The polysaccharides, also referred as holocellulose, constitute the most representative portion of bark components, accounting for more than a half of the bark weight. It is defined as water-insoluble carbohydrate fraction of plant materials mainly composed of cellulose, hemicellulose, and pectic substances (Heredia et al. 1995).

In the particular case of *E. globulus* bark, the carbohydrate fraction reported is about 62–79% (Table 1.3), being 43–56% cellulose (Miranda et al. 2012; Pereira 1988; Vázquez et al. 2008) and 20–24% hemicellulose (Miranda et al. 2012; Pereira 1988).

Cellulose defines the cell wall framework and is a three-dimensional linear homopolymer of poly-β (1,4)-D-glucose with a degree of polymerization greater than 15,000 (Heredia et al. 1995). It is mainly composed of a crystalline region with

Table 1.3 Chemical composition of *E. globulus* bark and extraction yields obtained in different solvents, expressed as % of total dry mass of bark

Parameter	%w/w$_{dried\ bark}$	Reference
Structural components		
Klason lignin	15.0–26.6	Miranda et al. (2013), Miranda et al. (2012), Mota et al. (2013), Neiva et al. (2014), Pereira (1988), and Vázquez et al. (2008)
Soluble lignin	1.4–7.5	Miranda et al. (2013), Mota et al. (2013), Neiva et al. (2014), Pereira (1988), and Vázquez et al. (2008)
Polysaccharides	62.5–79.7	Miranda et al. (2013), Miranda et al. (2012), Mota et al. (2013), Neiva et al. (2014), Pereira (1988), and Vázquez et al. (2008)
Suberin	0.98	Miranda et al. (2013)
Nonstructural components		
Ash content	1.6–4.7	Mota et al. (2013), Pereira (1988), and Vázquez et al. (2008)
Total extractives	6.0–8.5[a]	Miranda et al. (2013), Miranda et al. (2012), Neiva et al. (2014), and Pereira (1988)
Yields of extraction		
n-Hexane	0.42	Vázquez et al. (2008)
Benzene-ether	1.8–2.9	Pereira (1988)
Ether	0.68–1.3	Conde et al. (1996)
Dichloromethane	0.71–0.90[b]; 0.45[c]; 3.9[d]; 5.2[e]; 18.2[f]	Domingues et al. (2009), Freire et al. (2002a), and Mota et al. (2013)
Ethyl acetate	0.70	Vázquez et al. (2008)
Acetone	1.0	Vázquez et al. (2008)
Methanol	2.7–9.3 (boiling T) 8.2 (room T)	Mota et al. (2013), Santos et al. (2011), and Vázquez et al. (2008)
Ethanol	1.3–2.5	Vázquez et al. (2008)
Ethanol/toluene	2.2	Mota et al. (2013)
Methanol/water 80:20%v/v	2.0 2.8–12.3	Cadahía et al. (1997), Conde et al. (1996), Conde et al. (1995), and Vázquez et al. (2008)
Methanol/water 50:50%v/v	5.2	Vázquez et al. (2008)
Ethanol/water 80:20%v/v	2.6	Vázquez et al. (2008)
Ethanol/water 50:50%v/v	5.0 (boiling T) 9.3 (T ≈ 25 °C)	Vázquez et al. (2008) Santos et al. (2011)
2.5% Na$_2$SO$_3$ aqueous solution	8.5	Vázquez et al. (2008)
NaOH 0.1%	16.0	Mota (2011)
NaOH 1%	19.9–43.3	Miranda et al. (2013), Miranda et al. (2012), Mota et al. (2013), Pereira (1988), and Vázquez et al. (2008)

(continued)

Table 1.3 (continued)

Parameter	%w/w$_{dried\ bark}$	Reference
Water	4.1–8.1 (90 °C or boiling T)	Miranda et al. (2013), Miranda et al. (2012), Mota et al. (2013), and Vázquez et al. (2008)
	2.4–2.6 (room T/25 °C)	Mota et al. (2013) and Vázquez et al. (2008)

(a) Usually obtained with sequential extractions of several solvents, starting with dichloromethane, the less polar solvent that gives the nonpolar fraction of the extractives and followed by methanol, ethanol, and/or water (giving the polar fraction of bark extractives); dichloromethane extractives obtained in (b) trunk bark, (c) inner bark, (d) outer bark, (e) surface layers from trunk bark, and (f) branches bark

well-arranged cellulose macromolecules and contains a small portion of an amorphous region with the molecular chains arranged more irregularly. Cellulose is bonded with lignin and hemicellulose mainly by hydrogen bonding (Chen 2014). Being cellulose the most representative carbohydrate fraction, glucose is the main sugar moiety obtained after hydrolysis. In the particular case of *E. globulus* bark, a glucose range of 43–68% has been quantified after acid hydrolysis (Miranda et al. 2013; Mota et al. 2013; Vázquez et al. 2008). This value could be somewhat over estimated by a small contribution of glucose coming from glucomannans and non-cellulosic glucans.

Hemicellulose forms a network with cellulose microfibrils, and unlike cellulose, it is a heteropolymer with a small crystalline region composed of different monosaccharides, the hexoses (glucose, mannose, galactose, rhamnose), and the pentoses (xylose, arabinose). Additionally, some hemicelluloses contain uronic acids (Fengel and Wegener 1984).

It is expected that *E. globulus* bark hemicellulose has a similar structure to its respective wood containing glucuronoxylans and galacturanans (Pinto et al. 2005b; Shatalov et al. 1999). *E. globulus* wood xylans may contain 4-O-methyl-α-D-glucuronic acid or 4-O-methyl-α-D-glucuronic acid substituted with galactose or glucose linked with the xylan backbone. The substituted uronic acids can establish linking points between xylan and other cell wall polysaccharides such as rhamno-arabinogalactans and glucans (Pinto et al. 2005b). This is supported by the sugar moieties quantified after conventional acid hydrolysis and methanolysis of bark. Xylose is the second most abundant sugar moiety quantified either by conventional acid hydrolysis or methanolysis, accounting for 10.4–23.2% (Miranda et al. 2013; Mota et al. 2013; Vázquez et al. 2008). Minor amounts of arabinose (1.6–3.7%), galactose (1.6–3.6%), mannose (0.42–1.9%), and rhamnose (≈0.4%) have been quantified for *E. globulus* bark employing both methodologies (Miranda et al. 2013; Mota et al. 2013; Vázquez et al. 2008). Additionally, galacturonic (3.2%) and 4-O-methylglucuronic (2.3%) acids were only detected by acid methanolysis, explained by the fact that the labile uronic acids are easily degraded during conventional acid hydrolysis.

In this regard, Mota et al. (2013) have demonstrated that acid hydrolysis and acid methanolysis are methodologies that complement each other, giving a more reliable quantification of the sugar residues present. Due to its low degree of polymerization and crystalline structures, the hemicellulose is more easily degraded in acid medium impairing the respective quantification of the sugar residues, as is the case of the uronic acids not detected by conventional acid hydrolysis. On the other hand, acid methanolysis is unable to completely hydrolyze the cellulose into glucose.

Suberin
E. globulus bark contains 0.98% suberin confirming that its anatomical structure has a low amount of suberized phellem tissue in the periderm.

1.6.1.2   Nonstructural Components

Ash Content
The inorganic fraction of the bark is fundamental for cellular activity, and it refers to the ash content mainly composed of calcium, potassium, and magnesium originally in the form of carbonates, silicates, phosphates, and oxalates. Typically, ash content in *E. globulus* bark is below 5% (Table 1.3), a higher value than the respective wood (Fengel and Wegener 1984; Pereira 1988). Moreover, the inner bark ash content is higher than the outer bark.

In Miranda et al. (2013), the ash content and respective elemental composition were determined for different granulometric fractions of *E. globulus* bark, and differences according to the fraction analyzed were analyzed. Ash content obtained ranged from 4.3% to 23.1%, with an average value (12.1%) considerably higher than the values found in literature. The authors have explained these results with probable contamination of the soil and other extraneous fine materials during sample collection, pointing out the importance of bark cleaning during field and mill handling if a bark valorization step is to be considered. Elemental analysis has shown the presence of nitrogen (0.19–0.26%), calcium (0.181–0.623%), magnesium (0.120–0.154%), sodium (0.07–0.08%), and potassium (0.23–0.31%), among other elements (P, Cu, Zn, Ni, Cr, and Pb).

Extractives
Barks are usually richer in extractives than its corresponding wood, explained with the anatomical features of the bark and with the presence of the kino pockets or veins and oil glands (Eyles et al. 2004; Eyles et al. 2003; Hillis and Yazaki 1974). Kino pocket formation is associated with the tree mechanism of defense producing kino, a polyphenolic exudate containing several hydrophilic extractables such as polymeric proanthocyanidins, sometimes in their heterosidic form (Hillis and Yazaki 1974).

The extractives correspond to a small portion of the *E. globulus* bark (6.0–8.5%, Table 1.3) but are composed of a wide and complex variety of low molecular weight compounds ranging from lipophilic compounds (fat acids, aliphatic alcohols, triterpenes, and sterols), volatile terpenes, free sugars, and phenolic compounds (phenolic acids and aldehydes, cinnamic acids, stilbenes, hydrolysable, and condensed tannins, among others) (Cadahía et al. 1997; Conde et al. 1996; Conde et al. 1995; Domingues et al. 2009; Eyles et al. 2004; Freire et al. 2002b; Santos et al. 1997; Santos et al. 2011).

A significant portion of the *E. globulus* bark extractives are mainly of polar nature, corresponding to about 80–86% of the total extractives (Miranda et al. 2013; Miranda et al. 2012; Neiva et al. 2014), and are obtained with polar solvents water, methanol, and/or ethanol. The remaining compounds are mostly lipophilic compounds being extracted with less polar solvents such as dichloromethane.

As mentioned above, the age of the tree and environmental, genetic, and seasonal variations are also responsible for the wide intervals of extraction yields obtained among different studies employing the same solvent (Cadahía et al. 1997; Conde et al. 1996; Conde et al. 1995). Likewise, the methodologies used are also detrimental for the extraction yield. The majority of the extractions in bark chemical characterization are performed using the Soxhlet method, at the boiling temperature of the solvent used. In Table 1.3, literature data mostly concern to extractions performed at 90 °C or boiling temperature of the solvent. Some exceptions are indicated in the table, and extraction yields obtained at room temperature are also indicated. In Sect. 4.2, yields of extraction obtained with water, alkaline, and ethanol/water mixtures will be discussed with more detail and the effect of temperature and the solid-liquid ratio used addressed.

Existing literature displays a wide variety of extraction yields from *E. globulus* bark according to the type and polarity of the solvent. Since the majority of the extractives present in bark are polar, it is expected that extraction yields are higher in more polar solvents once it will remove the more hydrophilic components such as polyphenols, salts, and carbohydrates of low molecular weight (Rowell 2005). In this way, according to literature data summarized in Table 1.3, less polar solvents such as n-hexane, ether, ethyl acetate, or acetone have lower extraction yields between 0.42% and 1.3%. On the other hand, extractions performed with methanol, ethanol, water, and mixtures thereof have displayed a higher extraction yield interval, between 1.3% and 9.3%.

Literature data regarding the dichloromethane yield of extraction varies considerably because different fractions of bark have been studied (trunk, branches, surface layer, or peeling). Through Table 1.3 it is possible to observe that the lipophilic content of total trunk bark yields between 0.7% and 0.9% (Domingues et al. 2009; Mota et al. 2012). On the other hand, the outer layer of bark displays higher amount of lipophilic compounds (3.9%), and, if only the surface layers are considered, the yield obtained is higher (5.2%), and bark obtained from branches register the highest yield (18%) (Domingues et al. 2010). The inner bark has the lowest extraction yields in dichloromethane (0.45%). Moreover, it was also demonstrated in Domingues et al. (2010) study that the point of bark collection in the debarking process (e.g., wood yard stocks, pre-debarking, rotating debarking drum, bark grinding, bark accumulations over the bark grinder, and ground bark conveyer) is detrimental for the lipophilic extraction yield, obtaining a range interval of 0.6–1.5% of lipophilic compounds.

Extraction yields obtained with water, methanol, and ethanol can achieve a maximum of 8.1%, 9.3%, and 2.5%, respectively (Table 1.3). Studies encompassing the use of aqueous methanol/ethanol summarized in Table 1.3 have demonstrated that increasing the water content from 20% to 50%, the extraction yields increases up to two times (Vázquez et al. 2008).

Water extractions performed for temperatures above 90 °C have higher extraction yields (4.1–8.1%) than for extractions performed at room temperature (2.4–2.6%). This is explained by the fact that higher temperatures favor the solubility of compounds such as tannins and starch (Rowell 2005).

Alkaline extraction is commonly used to extract tannins from barks, offering a good alternative to organic solvents. It allows the extraction of flavonoid oligomers and polymers, waxes, suberin degradation products, carbohydrates, and lignin. Some existing literature is summarized in Table 1.3, and in Subsection 4.1, the yields and selectivity for extracting certain families of compounds will be discussed in more detail.

Through Table 1.3 it is possible to observe that the extractions performed with 2.5% Na_2SO_3 aqueous solution solubilize 8.5% of compounds, corresponding to half the content solubilized using NaOH 0.1 M (16%). Higher extraction yields ranging from 19.9% to 43.3% are obtained when a more concentrated solution of NaOH (1 M) is employed.

1.6.2   Current and Potential Commercial Products from the Bark

Nowadays, companies worldwide are concerned about developing environmental suitable processes in order to meet the increasing energy demand and, thus, efficiently convert biomass into several profitable and marketable commodities.

In the particular case of *E. globulus*, large amounts of bark are generated from the pulp and paper industry in Southern Europe. This by-product is currently used as fuel to suppress the power needs at the mill site. This is one of the simplest ways to deal with this high available resource since the complexity of managing with the physical and chemical composition heterogeneity of barks can be disregarded. On the other hand, it is a low-grade application for bark considering the low-value return and the operational problems that may result from the combustion stage caused by fouling problems triggered by the high ash content typically present in barks (Feng et al. 2013).

Bark could be a cheaper and important fiber source to suppress the needs of the pulp mills; however, it is not employed in pulp production since it lowers the quality and productivity of cellulose pulp, mainly attributed to the presence of high content of extractives and ashes that lowers the pulp yields (Amidon 1981), and it demands high consumption of cooking chemicals to achieve acceptable kappa numbers and leads to pulps with low brightness (Miranda et al. 2012; Neiva et al. 2014).

There are some studies considering the incorporation of a certain amount of cellulose pulp from *E. globulus* bark. Miranda and co-workers (Miranda et al. 2012) have demonstrated that although cellulose pulp from *E. globulus* bark is obtained with lower yields, delignification degree, and strength, the refining step is simpler than the corresponding wood pulp. The authors have demonstrated that bark can be incorporated in wood pulping up to 14% without significantly affecting pulp yield, delignification degree, refining, and strength properties (tensile index, tear index, and burst index). Nevertheless, it has to be counterbalanced the increase of organic material solubilized in the black liquor that needs to be processed and the environmental issues related with the effluents resulting from bleaching process.

Biochar, a carbon-rich material, can be produced from bark pyrolysis and employed as fuel, in the preparation of activated carbons and soil amendment (Mohan et al. 2006). In this thermochemical conversion process, it is possible to simultaneously produce bio-oil and syngas for further valorization into energy and as feedstock for chemical production. Other thermochemical conversion process typically employed with forestry biomass residues is the liquefaction at high temperatures in the presence of water or any other solvent (e.g., ethanol, acetone, or phenol) for the production of energy, valuable chemicals, or composite materials such as resins, polyurethanes, or polyesters. A resin-type molding material could be envisaged for wood floor fillers and conventional molding agents: bark-based thermosetting materials were obtained after *Eucalyptus* bark phenolysis in the presence of sulfuric acid being very similar to the commercially novalak resin-based molding materials (Alma and Kelley 2000; Zhao et al. 2006). Torrefaction of *E. globulus* bark has been studied for pelletization (Arteaga-Pérez et al. 2015). The thermochemical conversion processes applied to different biomasses, including bark, have been extensively reviewed (Canabarro et al. 2013; Feng et al. 2013; Gollakota et al. 2018; Kumar et al. 2018; Tanger et al. 2013; Verma et al. 2012).

Being *E. globulus* bark a plentiful available forest biomass resource well suitable to obtain lignin, cellulose, hemicellulose, and extractives, within the biorefinery concept applied to pulp and paper industry, there are considerable research efforts in finding upgraded valorization applications (de Melo et al. 2012; Domingues et al. 2010; Pinto et al. 2013; Pinto et al. 2017; Santos et al. 2011; Vankar et al. 2009; Vázquez et al. 2009). In Feng et al. (2013), the potential of converting bark from several species into materials and extract high value-added chemicals is discussed, highlighting the main breakthroughs achieved and obstacles to overcome in the future. Several studies conducted with barks of different species have shown the great potential of this biomass to prepare adsorbents, fillers for phenolic resins to be applied in plywood, or as a source of bioactive compounds, among many other applications. Figure 1.18 depicts an overview of the possible valorization routes for *E. globulus* bark, and although some of them were never demonstrated for *E. globulus* bark, they have been successfully explored with other barks. Identifying the most promising valorization routes has been a challenge through the years due to the heterogeneous structure of barks, and developing a suitable industrial process still requires more research to proper explore the most profitable one.

In the integrated biorefinery context, the exploitation of bark extractives previously to its current use of burning to obtain energy constitutes a promising approach. It can include the previous extraction of the lipophilic compounds before extracting the more polar compounds with water, methanol, ethanol, or alkaline solutions. Although this type of compounds has high value in the market, the development of an industrial process probably should consider the separation of several compounds simultaneously and its purification in order to obtain an economically feasible process.

The low molecular weight compounds of several barks have been extensively studied and several lipophilic and phenolic compounds identified displaying interesting chemical, biological, and pharmacological properties (Babayi et al. 2004; Hou et al. 2000; Kampa et al. 2004; Meltzer and Malterud 1997; Moure et al. 2001; Perchellet et al. 1994; Pizzi 2008; Takuo 2005).

The polyphenolic compounds correspond to a significant portion of the polar extractables of *E. globulus* bark and will be detailed in Sect. 4. Several phenolic acids and aldehydes, flavonoids, flavonoid glycosides, and galli- and ellagi-tannins have already been identified in methanolic extracts (Santos et al. 2013). Many studies have demonstrated the potential of the polar extracts as antioxidants (Vázquez et al. 2009; Vázquez et al. 2012) and anticarcinogenic (Mota et al. 2012).

Regarding the less polar compounds, the main volatile terpenes identified after dichloromethane extraction of the *E. globulus* bark were aromadendrene, a-pinene, 1.8-cineole, a-phellandrene, and virididlorene (Eyles et al. 2004; Eyles et al. 2003). The major families of lipophilic compounds identified in *E. globulus* bark correspond to fatty acids, long-chain aliphatic alcohols, triterpenoids, and sterols (Freire et al. 2002b). Lipophilic extractives of inner and outer bark differ significantly in

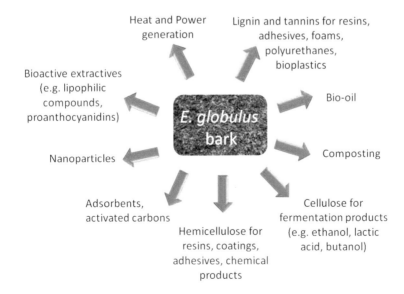

Fig. 1.18 Possible valorization routes for *E. globulus* bark

composition. The inner bark lipophilic compounds are very close to the respective wood, and for the outer bark, the lipophilic extractive content is mostly represented by triterpenic acids with lupine, ursane, and oleanane skeletons. Significant amounts of sesquiterpenes and sesquiterpene alcohols have also been identified in outer bark. Fatty acids and alcohols were found in both fractions. *E. globulus* bark potential for obtaining bioactive lipophilic compounds has also been demonstrated (Domingues et al. 2010; Freire et al. 2002b), and recently, Domingues et al. (2012) managed to obtain concentrated fractions of several triterpenoid acids with confirmed bioactivity, such as ursolic, betulinic, oleanolic, butulonic, 3-acetylursolic, and 3-acetyloleanolic acids, by a two-stage supercritical fluid chromatographic process employing soft conditions in the first stage (120 bar, 40 °C, and pure CO_2 as the extracting agent) and a second stage with more harsh conditions (200 bar, 40 °C, and modified CO_2 extracting agent with 5% ethanol).

The production of nanoparticles in the development of nanomaterials for catalysis, sensors, and biomedical applications is an emergent trend. Aqueous extracts of *E. globulus* bark were employed in the green synthesis of metallic nanoparticles of silver and gold, and it was demonstrated that the polyphenolic galloyl derivatives present are predominantly responsible for the metal-ion reduction, while the sugar fraction and ellagic acid and isorhamnetin have the function of stabilizing the nanoparticles (Santos et al. 2014).

Tannins are natural occurring phenolic compounds and are the main family of compounds in bark extractives, easily extracted employing alkaline solutions. They are suitable clarification agents for beverages due to their capability of complexing metal ions. Crude fractions of bark tannins are typically employed in leather tannin industry and have a great potential to replace phenol compounds normally employed as binging agents in the production of plywoods and laminates, as flotation agents and as cement superplasticizers (Pinto et al. 2013; Pizzi 1998; USDA 1971). The suitability of alkaline *E. globulus* bark extracts for leather tannin has already been demonstrated (Pinto et al. 2013). Several studies have successfully developed bark tannin-based adhesives for wood board, particleboard, and cardboard production as well as tannin-resorcinol-formaldehyde cold-setting waterproof adhesives (Pizzi 2008). Moreover, tannins are natural precursors for biodegradable polyurethane and rigid foam formulations once they contain both aliphatic and aromatic hydroxyl groups (Celzard et al. 2010; Feng et al. 2013; Lacoste et al. 2015).

Lignin plays an important role in the plant being very well known for their antioxidant and antimicrobial activities (Lora and Glasser 2002). Typically, the lignin in bark is chemically very similar to the corresponding wood. The species and the processes applied to isolate and purify lignin will determine the possible end uses. Modification of lignin can also tailor properties for specific applications (Duval and Lawoko 2014).

Typically lignin is used as fuel source; however, similar to tannins, lignin can also be a suitable replacement for phenolic resins as binders and can be used in the formulation of polyurethane foams, epoxy resins, and lignin-based bio-dispersants (Lora and Glasser 2002). In Khan et al. (2004), lignin-based phenol formaldehyde adhesives up to 50%wt. demonstrated having better bonding strength than the phenol formaldehyde resins frequently employed and can be a good replacement for plywood production. Several efforts are being made in order to develop thermoplastic materials mainly employing lignin for several purposes (Laurichesse and Avérous 2014; Lora and Glasser 2002). Apart from the discussed applications, several works have shown the potential of lignin as raw material to produce carbon fibers and valuable phenolic compounds (Chang et al. 2014; Costa et al. 2014; Duval and Lawoko 2014).

Bark polysaccharides include cellulose and hemicellulose and could be explored for several industrial applications as, for example, in fine chemicals, cosmetics, and nutraceuticals. The production of alcohols, butanol, and mainly ethanol from polysaccharides is the most explored valorization route. Although bark is not a favorable source to obtain fermentable sugars due to the high content in extractives, a pretreatment could be envisaged in order to consider bark as a promising source for ethanol production. The production of other chemicals from polysaccharides is extensive ranging from furans, sugar alcohols, sugar acids with broad applications as adhesives, sweeteners and precursors for fine chemicals, and plastics, among many others (Moon et al. 2011).

Functional polymers and materials can be made out of cellulose and are extensively studied. Several different particles with distinct physical and chemical, mechanical, rheological, and thermal properties can be obtained ranging from micro- and nano-fibrillated cellulose, crystalline nanocellulose, and bacterial cellulose (Moon et al. 2011).

Bark valorization routes are extensive, being several of the approaches suggested here not yet reported for *E. globulus* bark. The choice of the most promising route(s) will be determined by the resources available at the mill site and most of all the return in investment.

References

Ahvazi B, Cloutier É, Wojciechowicz O, Ngo T-D (2016) Lignin profiling: a guide for selecting appropriate lignins as precursors in biomaterials development. ACS Sustain Chem Eng 4:5090–5105

Alma MH, Kelley SS (2000) Conversion of barks of several tree species into bakelite-like thermosetting materials by their phenolysis. J Polym Eng 20(5):365–380

Alriols MG, García A, Llano-Ponte R, Labidi J (2010) Combined organosolv and ultrafiltration lignocellulosic biorefinery process. Chem Eng J 157(1):113–120

American Process Inc (2015) https://americanprocess.com/bioplus/news/. Accessed 08 Feb 2018

American Science and Technology (2017) http://www.amsnt.com/organosolv/. Accessed 08 Feb 2018

Amidon TE (1981) Effect of wood properties of hardwoods on kraft paper properties. TAPPI J 64:123–126

Argyropoulos DS, Abacherli A, Rincón AG, Arx UV (2009) Quantitative ^{31}P nuclear magnetic resonance (NMR) spectra of lignin. International Lignin Institute, Lausanne, Switzerland

Arteaga-Pérez LE, Segura C, Bustamante-García V, Jiménez R (2015) Torrefaction of wood and bark from *Eucalyptus globulus* and *Eucalyptus nitens*: focus on volatile evolution vs feasible temperatures. Energy 93:1731–1741

AVAPCO (2011) Technology – A sugar platform. http://avapco.com/technology.html. Accessed 07 Feb 2018

Axegård P (2016) Lignin processing and lignin applications. In: BPM Lignin Satelite Workshop, Wageningen, June 17

Azadi P, Inderwildi OR, Farnood R, King DA (2013) Liquid fuels, hydrogen and chemicals from lignin: a critical review. Renew Sust Energ Rev 21:506–523

Babayi H, Kolo I, Okogun JI, Ijah UJJ (2004) The antimicrobial activities of Methanolic extracts of *Eucalyptus camaldulensis* and *Terminalia catappa* against some pathogenic microorganisms. Biokemistri 16:106–111

Bagh FSG, Lalman J, Seth R, Ray S, Biswas N (2017) Using deep eutectic solvents (DESs) to extract lignin from black liquor. In: 17 AIChE annual meeting, Minneapolis, U.S.A., October 29– November 3

Barbash VA, Yaschenko OV, Shniruk OM (2017) Preparation and properties of nanocellulose from organosolv straw pulp. Nanoscale Res Lett 12:241

Barsberg S, Matousek P, Towrie M (2005) Structural analysis of lignin by resonance Raman spectroscopy. Macromol Biosci 5:743–752

Behling R, Valange S, Chatel G (2016) Heterogeneous catalytic oxidation for lignin valorization into valuable chemicals: what results? What limitations? What trends? Green Chem 18:1839

Berlin A, Balakshin M (2014) Chapter 18 - industrial lignins: analysis, properties, and applications. In: Gupta VK, Tuohy MG, Kubicek CP, Saddler J, Xu F (eds) Bioenergy research: advances and applications. Elsevier, Amsterdam, pp 315–336

Biofuels Digest (2014) Stora Enso acquires Virdia in (up to) $62M deal. http://www.biofuelsdigest.com/bdigest/2014/06/23/stora-enso-acquires-virdia-in-up-to-62m-deal/. Accessed 08 Feb 2018

Buranov AU, Mazza G (2008) Lignin in straw of herbaceous crops. Ind Crop Prod 28:237–259

Burgo Group (2013) Ligninsulphonates. http://www.burgo.com/en/group/figures/ls. Accessed 08 Feb 2018

Cadahía E, Conde E, Simon BF, GarciaVallejo MC (1997) Tannin composition of *Eucalyptus camaldulensis*, *E. globulus* and *E. rudis*. Part II. Bark. Holzforsch. - Int J Biol Chem Phys Technol Wood 51:125–129

Calvo-Flores FG, Dobado JA (2010) Lignin as renewable raw material. ChemSusChem 3:1227–1235

Canabarro N, Soares JF, Anchieta CG, Kelling CS, Mazutti MA (2013) Thermochemical processes for biofuels production from biomass. Sustainable Chem Processes 1:22

Capanema EA, Balakshin MY, Chen CL, Gratzl JS, Gracz H (2001) Structural analysis of residual and technical lignins by ^{1}H-^{13}C correlation 2D NMR-spectroscopy. Holzforsch. – Int J Biol Chem Phys Technol Wood 55:302–308

Capanema EA, Balakshin MY, Kadla JF (2004) A comprehensive approach for quantitative lignin characterization by NMR spectroscopy. J Agric Food Chem 52:1850–1860

Celzard A, Zhao W, Pizzi A, Fierro V (2010) Mechanical properties of tannin-based rigid foams undergoing compression. Mater Sci Eng A 527:4438–4446

CEPI (2017) Key statistics 2016. http://www.cepi.org/system/files/public/documents/publications/statistics/2017/KeyStatistics2016_Final.pdf. Accessed 08 Feb 2018

Chakar FS, Ragauskas AJ (2004) Review of current and future softwood kraft lignin process chemistry. Ind Crop Prod 20:131–141

Chang CF et al (2014) Carbon Fibers Derived from Lignin. Patent Application PCT/US2013/055654. https://patentimages.storage.googleapis.com/84/5f/56/1d18819ff9088a/WO2014046826A1.pdf

Chempolis (2017) http://www.chempolis.com/. Accessed 08 Feb 2018

Chen H (2014) Biotechnology of lignocellulose. Theory and practice. In: Springer (ed), Dordrecht, Netherlands

Clariant (2018) Clariant to build flagship sunliquid® cellulosic ethanol plant in Romania. https://www.clariant.com/en/Corporate/News/2017/10/Clariant-to-build-flagship-sunliquid-cellulosic-ethanol-plant-in-Romania. Accessed 09 Feb 2018

Clayton D et al (1983) Overview. In: Grace TM, Leopold B, Malcolm E, Kocurek MJ (eds) Pulp and paper manufacture. Alkaline pulping, vol 5. TAPPI/CPPA, Atlanta/Montreal, TAPPI - Atlanta; CPPA - Montreal, pp 3–14

Conde E, Cadahia E, Diez-Barra R, García-Vallejo MC (1996) Polyphenolic composition of bark extracts from *Eucalyptus camaldulensis*, *E. globulus* and *E. rudis*. Eur J Wood Wood Prod 54:175–181

Conde E, Cadahía E, García-Vallejo MC, Tomás-Barberán F (1995) Low molecular weight polyphenols in wood and bark of *Eucalyptus globulus*. Wood Fiber Sci 27:379–383

Constant S et al (2016) New insights into the structure and composition of technical lignins: a comparative characterisation study. Green Chem 18:2651–2665

Costa CAE, Coleman W, Dube M, Rodrigues AE, Pinto PCR (2016) Assessment of key features of lignin from lignocellulosic crops: stalks and roots of corn, cotton, sugarcane, and tobacco. Ind Crop Prod 92:136–148

Costa CAE, Pinto PCR, Rodrigues AE (2014) Evaluation of chemical processing impact on *E. globulus* wood lignin and comparison with bark lignin. Ind Crop Prod 61:479–491

Costa CAE, Pinto PCR, Rodrigues AE (2015) Radar tool for lignin classification on the perspective of its valorization. Ind Eng Chem Res 54:7580–7590

Costa CAE, Pinto PCR, Rodrigues AE (2018) Lignin fractionation from *E. Globulus* kraft liquor by ultrafiltration in a three stage membrane sequence. Sep Purif Technol 192:140–151

de Melo MMR, Oliveira ELG, Silvestre AJD, Silva CM (2012) Supercritical fluid extraction of triterpenic acids from *Eucalyptus globulus* bark. J Supercrit Fluids 70:137–145

Dale B (2018) Time to rethink cellulosic biofuels? Biofuels Bioprod Biorefin 12:5–7

Domingues R, Sousa G, Freire C, Silvestre A, Neto C (2009) Eucalyptus globulus biomass residues from pulping industry as a source of high value triterpenic compounds. Ind Crop Prod 31:65–70

Domingues RMA et al (2012) Supercritical fluid extraction of *Eucalyptus globulus* bark-a promising approach for triterpenoid production. Int J Mol Sci 13:7648–7662

Domingues RMA, Sousa GDA, Freire CSR, Silvestre AJD, Neto CP (2010) *Eucalyptus globulus* biomass residues from pulping industry as a source of high value triterpenic compounds. Ind Crop Prod 31:65–70

Domtar (2013) BioChoice™ lignin. http://www.upmbiochemicals.com/SiteCollectionDocuments/BioChoice-brochure.pdf. Accessed 15 Feb 2018

Duval A, Lawoko M (2014) A review on lignin-based polymeric, micro- and nano-structured materials. React Funct Polym 85:78–96

El Hage R, Brosse N, Chrusciel L, Sanchez C, Sannigrahi P, Ragauskas A (2009) Characterization of milled wood lignin and ethanol organosolv lignin from *miscanthus*. Polym Degrad Stab 94:1632–1638

El Hage R, Brosse N, Sannigrahi P, Ragauskas A (2010) Effects of process severity on the chemical structure of *Miscanthus* ethanol organosolv lignin. Polym Degrad Stab 95:997–1003

Evtuguin DV, Neto CP, Silva AMS, Domingues PM, Amado FML, Robert D, Faix O (2001) Comprehensive study on the chemical structure of dioxane lignin from plantation *Eucalyptus globulus* wood. J Agric Food Chem 49:4252–4261

Eyles A, Davies N, Mohammed C (2004) Traumatic oil glands induced by pruning in the wound-associated phloem of *Eucalyptus globulus*: chemistry and histology. Trees: Struct Funct 18:204–210

Eyles A, Davies N, Yuan Z, Mohammed C (2003) Host responses to natural infection by *Cytonaema sp.* in the aerial bark of *Eucalyptus globulus*. For Pathol 33:317–331

Ek M, Gellerstedt, G, Henriksson (2009) Pulp and Paper Chemistry and Technology, Pulping Chemistry and Technology, vol 2. de Gruyter, Berlin

Faix O, Argyropoulos Dimitris S, Robert D, Neirinck V (1994) Determination of hydroxyl groups in lignins evaluation of ^{1}H, ^{13}C, ^{31}P NMR, FTIR and wet chemical methods. Holzforsch. – Int J Biol Chem Phys Technol Wood 48:387

Feng S, Cheng S, Yuan Z, Leitch M, Xu C (2013) Valorization of bark for chemicals and materials: a review. Renew Sust Energ Rev 26:560–578

Fengel D, Wegener G (1984) Lignin. In: Wood: chemistry, ultrastructure, reactions. Walter de Gruyter, New york, pp 132–181

Fernández-Costas C, Gouveia S, Sanromán MA, Moldes D (2014) Structural characterization of Kraft lignins from different spent cooking liquors by 1D and 2D nuclear magnetic resonance spectroscopy. Biomass Bioenergy 63:156–166

Fraser W (2018) Lignin. https://www.westfraser.com/products/lignin-0. Accessed 15 Feb 2018

Freire C, Silvestre A, Neto C (2002a) Identification of new hydroxy fatty acids and ferulic acid esters in the wood of *Eucalyptus globulus*. Holzforsch. – Int J Biol Chem Phys Technol Wood 56:143–149

Freire C, Silvestre A, Neto C, Cavaleiro J (2002b) Lipophilic extractives of the inner and outer barks of *Eucalyptus globulus*. Holzforsch. – Int J Biol Chem Phys Technol Wood 56:372–379

Froass PM, Ragauskas AJ, Jiang J (1998) Nuclear magnetic resonance studies. 4. Analysis of residual lignin after kraft pulping. Ind Eng Chem Res 37:3388–3394

Gellerstedt G, Henriksson G (2008) Lignins: major sources, structure and properties. In: Gandini A (ed) Monomers, polymers and composites from renewable resources. Elsevier, Amsterdam, pp 201–224

Gollakota ARK, Kishore N, Gu S (2018) A review on hydrothermal liquefaction of biomass. Renew Sust Energ Rev 81:1378–1392

Gooding C (2012) Comparison of the LignoBoost and SLRP lignin recovery process. In: Interbational Bioenergy & Bioproducts Conference (IBBC), Savannah, Georgia, October 17–19

Gordobil O, Moriana R, Zhang L, Labidi J, Sevastyanova O (2016) Assesment of technical lignins for uses in biofuels and biomaterials: structure-related properties, proximate analysis and chemical modification. Ind Crop Prod 83:155–165

Gosselink R (2011) Lignin as a renewable aromatic resource for the chemical industry. In: Mini-symposium organised by Wageningen UR lignin platform, Wageningen, December 6

Gullichsen J (1999) Fiber line operations. In: Gullichsen J, Fogelholm CJ (eds) Chemical pulping, Papermaking science and technology, vol 6A. Fapet Oy, Jyväskylä, Finland, pp 19–243

Hallac BB, Pu Y, Ragauskas AJ (2010) Chemical transformations of *Buddleja davidii* lignin during ethanol organosolv pretreatment. Energy Fuel 24:2723–2732

Harlow S (2016) Renmatix turns biomass into sugars for industrial use. articles.extension.org/pages/73638/renmatix-turnsbiomass-into-sugars-for-industrial-use. Accessed 08 Feb 2018

Harkin, JM, Rowe JW (1971) Bark And Its Possible Uses. Research note, U.S. Department of Agriculture, Forest Service, Forest Products Laboratory, Research note 091, Madison, U.S.A.

Heitner C, Dimmel D, Schmidt J (2010) Lignin and lignans: advances in chemistry. CRC Press, Taylor & Francis, Florida, U.S.A.

Heredia AJ, Jiménez A, Guillén R (1995) Composition of plant cell walls. Z Lebensm Unters Forsch 200:24–31

Hillis W, Yazaki Y (1974) Kinos of Eucalyptus species and their acid degradation products. Phytochemistry 13:495–498

Hou A-J, Liu Y-Z, Yang H, Lin Z-W, Sun H-D (2000) Hydrolyzable tannins and related polyphenols from *Eucalyptus globulus*. J Asian Nat Prod Res 2:205–212

Huang C, He J, Narron R, Wang Y, Yong Q (2017) Characterization of kraft lignin fractions obtained by sequential ultrafiltration and their potential application as a biobased component in blends with polyethylene. ACS Sustain Chem Eng 5:11770–11779

Humpert D, Ebrahimi M, Czermak P (2016) Membrane technology for the recovery of lignin: a review. Membranes 6:42

Ibarra D et al (2007) Lignin modification during *Eucalyptus globulus* kraft pulping followed by totally chlorine-free bleaching: a two-dimensional nuclear magnetic resonance, fourier transform infrared, and pyrolysis–gas chromatography/mass spectrometry study. J Agric Food Chem 55:3477–3490

Iogen Corporation (2015) Costa pinto project. https://www.iogen.ca/raizen-project/index.html. Accessed 09 Feb 2018

Jönsson J, Berntsson T 2010 Analysing the potential for CCS within the European pulp and paper industry. In: 23rd international ECOS conference, Lausanne, Switzerland, June 14–17, pp 676–683

Kampa M et al (2004) Antiproliferative and apoptotic effects of selective phenolic acids on T47D human breast cancer cells: potential mechanisms of action. Breast Cancer Res 6:R63–R74

Kautto J (2017) Evaluation of two pulping-based biorefinery concepts. Lappeenranta University of Technology. https://www.doria.fi/bitstream/handle/10024/143661/Jesse%20Kautto%20A4_ei%20artik_1.pdf?sequence=2. Accessed 08 Feb 2017

Kautto J, Realff MJ, Ragauskas AJ (2013) Design and simulation of an organosolv process for bioethanol production. Biomass Convers Bior 3:199–212

Khan MA, Ashraf SM, Malhotra VP (2004) Eucalyptus bark lignin substituted phenol formaldehyde adhesives: a study on optimization of reaction parameters and characterization. J Appl Polym Sci 92:3514–3523

Kim J-P, Lee I-K, Yun B-S, Chung S-H, Shim G-S, Koshino H, Yoo I-D (2001) Ellagic acid rhamnosides from the stem bark of *Eucalyptus globulus*. Phytochemistry 57:587–591

Kouisni L, Gagné A, Maki K, Holt-Hindle P, Paleologou M (2016) LignoForce system for the recovery of lignin from black liquor: feedstock options, odor profile, and product characterization. ACS Sustain Chem Eng 4(10):5152–5159

Kouisni L, Holt-Hindle P, Maki K, Paleologou M (2012) The Lignoforce system™: a new process for the production of high-quality lignin from black liquor. J Sci Technol Prod Process 2:6–10

Kumar AK, Sharma S (2017) Recent updates on different methods of pretreatment of lignocellulosic feedstocks: a review. Bioresour Bioprocess 4:7

Kumar M, Olajire Oyedun A, Kumar A (2018) A review on the current status of various hydrothermal technologies on biomass feedstock. Renew Sust Energ Rev 81:1742–1770

Kun D, Pukánszky B (2017) Polymer/lignin blends: interactions, properties, applications. Eur Polym J 93:618–641

Lacoste C, Basso MC, Pizzi A, Celzard A, Ella Ebang E, Gallon N, Charrier B (2015) Pine (*P. pinaster*) and quebracho (*S. lorentzii*) tannin-based foams as green acoustic absorbers. Ind Crop Prod 67:70–73

Lake MA, Blackburn JC (2011) Process for recovering lignin, US Patent Application 2011/0294991

Lake MA, Blackburn JC (2014) SLRP™ – an innovative lignin-recovery technology. Cellul Chem Technol 48(9–10):799–804

Lange H, Schiffels P, Sette M, Sevastyanova O, Crestini C (2016) Fractional precipitation of wheat straw Organosolv lignin: macroscopic properties and structural insights. ACS Sustain Chem Eng 4:5136–5151

Laurichesse S, Avérous L (2014) Chemical modification of lignins: towards biobased polymers. Prog Polym Sci 39:1266–1290

Lin SY, Dence CW (1992) Methods in lignin chemistry. Springer-Verlag, Berlin, Germany

Lora J (2008) Industrial commercial lignins: sources, properties and applications. In: Belgacem MN, Gandini A (eds) Monomers, polymers and composites from renewable resources. Elsevier, Amsterdam, pp 225–241

Lora JH, Glasser WG (2002) Recent industrial applications of lignin: a sustainable alternative to nonrenewable materials. J Polym Environ 10:39–48

Lourenço A, Gominho J, Marques AV, Pereira H (2012) Variation of lignin monomeric composition curing kraft pulping of *Eucalyptus globulus* heartwood and sapwood. J Wood Chem Technol 33:1–18

Lu F, Ralph J (1997) Derivatization followed by reductive cleavage (DFRC method), a new method for lignin analysis: protocol for analysis of DFRC monomers. J Agric Food Chem 45:2590–2592

Mahmood N, Yuan Z, Schmidt J, Xu C (2016) Depolymerization of lignins and their applications for the preparation of polyols and rigid polyurethane foams: a review. Renew Sust Energ Rev 60:317–329

Mansouri NE, Salvadó J (2006) Structural characterization of technical lignins for the production of adhesives: application to lignosulfonate, kraft, soda-anthraquinone, organosolv and ethanol process lignins. Ind Crop Prod 24:8–16

Mansouri NE, Salvadó J (2007) Analytical methods for determining functional groups in various technical lignins. Ind Crop Prod 26:116–124

McDonough TJ (1992) The chemistry of organosolv delignification. TAPPI J 76:186–193

Meltzer H, Malterud K (1997) Can dietary flavonoids influence the development of coronary heart disease? Scand J Nutr 41:50–57

Miller J, Faleiros M, Pilla L, Bodart AC (2016) Lignin: technology, applications and markets. Special market analysis study. RISI, Market-Intell LCC

Minor JL (1996) Raw materials. Production of unbleached pulp. In: Dence CW, Reeve DW (eds) Pulp bleaching – principles and practice. Tappi Press, Atlanta, pp 27–57

Miranda I, Gominho J, Mirra I, Pereira H (2013) Fractioning and chemical characterization of barks of *Betula pendula* and *Eucalyptus globulus*. Ind Crop Prod 41:299–305

Miranda I, Gominho J, Pereira H (2012) Incorporation of bark and tops in *Eucalyptus globulus* wood pulping. Bioresources 7:4350–4361

Mirra IMP (2011) Influência das diferentes granulometrias na composição química das cascas de *Eucalyptus globulus Labill., Betula pendula Roth, Picea abies (L.) Karst, Pinus sylvestris L. e Pinus pinea L..* Master thesis, Universidade Técnica de Lisboa

Mohan D, Pittman CU, Steele PH (2006) Pyrolysis of wood/biomass for bio-oil: a critical review. Energy Fuel 20:848–889

Monteil-Rivera F, Phuong M, Ye M, Halasz A, Hawari J (2013) Isolation and characterization of herbaceous lignins for applications in biomaterials. Ind Crop Prod 41:356–364

Moon RJ, Martini A, Nairn J, Simonsen J, Youngblood J (2011) Cellulose nanomaterials review: structure, properties and nanocomposites. Chem Soc Rev 40:3941–3994

Mota I (2011) Extracção de base aquosa de compostos polares da casca de *Eucalyptus globulus* na perspectiva da sua recuperação. Master thesis, Faculty of Engeneering of University of Porto

Mota I et al (2012) Extraction of polyphenolic compounds from *Eucalyptus globulus* bark: process optimization and screening for biological activity. Ind Eng Chem Res 51:6991–7000

Mota I et al (2013) *Eucalyptus globulus* bark as a source of polyphenolic compounds with biological activity. O Papel 74:57–64

Moure A et al (2001) Natural antioxidants from residual sources. Food Chem 72:145–171

Myerly RC, Nicholson MD, Katzen R, Taylor JM (1981) The forest refinery. ChemTech 11:186–192

Neiva DM, Gominho J, Pereira H (2014) Modeling and optimization of *Eucalyptus globulus* bark and wood delignification using response surface methodology. BioResources 9:2907–2921

Nimz HH, Robert D, Faix O, Nemr M (1981) Carbon-[13]NMR spectra of lignins, 8. Structural differences between lignins of hardwoods, softwoods, grasses and compression wood. Holzforsch. – Int J Biol Chem Phys Technol Wood 35:16–26

Oliveira RCP, Mateus M, Santos DMF (2016) Black liquor electrolysis for hydrogen and lignin extraction. ECS Trans 72:43–53

Pan X et al (2005) Biorefining of softwoods using ethanol organosolv pulping: preliminary evaluation of process streams for manufacture of fuel-grade ethanol and co-products. Biotechnol Bioeng 90:473–481

Pan X et al (2006) Bioconversion of hybrid poplar to ethanol and co-products using an organosolv fractionation process: optimization of process yields. Biotechnol Bioeng 94:851–861

Panagiotopoulos IA, Chandra RP, Saddler JN (2012) A two-stage pretreatment approach to maximise sugar yield and enhance reactive lignin recovery from poplar wood chips. Bioresour Technol 130:570–577

Pepper JM, Casselman BW, Karapally JC (1967) Lignin oxidation. Preferential use of cupric oxide. Can J Chem 45:3009–3012

Perchellet JP, Gali HU, Perchellet EM, Laks PE, Bottari V, Hemingway RW, Scalbert A (1994) Antitumor-promoting effects of gallotannins, ellagitannins, and flavonoids in mouse skin in vivo. In: Huang M-T, Osawa T, Ho C-T, Rosen RT (eds) Food phytochemicals for Cancer prevention I: fruits and vegetables, vol 546. ACS symposium series. American Chemical Society, Washington, DC, pp 303–327

Pereira H (1988) Variability in the chemical composition of plantation eucalyptus (*Eucalyptus globulus*), Wood Fiber Sci 20(1):82–90

Picardo MC, de Medeiros JL, Monteiro JGM, Chaloub RM, Giordano M, de Queiroz Fernandes Araújo O (2013) A methodology for screening of microalgae as a decision making tool for energy and green chemical process applications. Clean Techn Environ Policy 15:275–291

Pinto PC, Borges da Silva EA, Rodrigues AE (2011) Insights into oxidative conversion of lignin to high-added-value phenolic aldehydes. Ind Eng Chem Res 50:741–748

Pinto PC, Evtuguin DV, Neto CP (2005a) Effect of structural features of wood biopolymers on hardwood pulping and bleaching performance. Ind Eng Chem Res 44:9777–9784

Pinto PC, Evtuguin DV, Neto CP (2005b) Structure of hardwood glucuronoxylans: modifications and impact on pulp retention during wood kraft pulping. Carbohydr Polym 60:489–497

Pinto PC, Evtuguin DV, Neto CP, Silvestre AJD (2002a) Behavior of *Eucalyptus globulus* lignin during kraft pulping I. Analysis by chemical degradation methods. J Wood Chem Technol 22:93–108

Pinto PC, Evtuguin DV, Neto CP, Silvestre AJD, Amado FML (2002b) Behavior of *Eucalyptus globulus* lignin during kraft pulping II. Analysis by NMR, ESI/MS and GPC. J Wood Chem Technol 22:109–125

Pinto PCR, Borges da Silva EA, Rodrigues AE (2012) Lignin as source of fine chemicals: vanillin and syringaldehyde. In: Baskar C, Baskar S, Dhillon RS (eds) Biomass conversion. Springer, Berlin, Heidelberg, pp 381–420

Pinto PCR, Oliveira C, Costa CAE, Rodrigues AE (2016) Performance of side-streams from Eucalyptus processing as sources of polysaccharides and lignins by kraft delignification. Ind Eng Chem Res 55:516–526

Pinto PCR, Sousa G, Crispim F, Silvestre AJD, Neto CP (2013) *Eucalyptus globulus* bark as source of tannin extracts for application in leather industry. ACS Sustain Chem Eng 1:950–955

Pinto PR, Mota IF, Pereira CM, Ribeiro AM, Loureiro JM, Rodrigues AE (2017) Separation and recovery of polyphenols and carbohydrates from *Eucalyptus* bark extract by ultrafiltration/diafiltration and adsorption processes. Sep Purif Technol 183:96–105

Pizzi A (1998) Wood Bark Extracts as Adhesives and Preservatives. In: Bruce A, Palfreyman JW (eds) Forest Products Biotechnology. Taylor & Francis Ltd, London, United Kingdom

Pizzi A (2008) Tannins: major sources, properties and applications. In: Belgacem MN, Gandini A (eds) Monomers, polymers and composites from renewable resources, 1st edn. Elsevier, Oxford, pp 179–199

PLANT (2016) West Fraser brews its black liquor. https://www.plant.ca/features/west-fraser-brews-black-liquor/. Accessed 15 Feb 2018

POET-DSM Advanced Biofuels (2014) POET-DSM achieves cellulosic biofuel breakthrough. http://poetdsm.com/pr/poet-dsm-achieves-cellulosic-biofuel-breakthrough. Accessed 08 Feb 2018

Popa V (2013) Pulping fundamentals and processing. In: Popa V (ed) Pulp Production and processing: from papermaking to high-tech products. Smithers Rapra, Shropshire, UK, pp 35–70

Quesada J, Rubio M, Gomez D (1999) Ozonation of lignin rich solid fractions from corn stalks. J Wood Chem Technol 19:115–137

Quilhó T, Pereira H (2001) Within and between-tree variation of bark content and wood density of *Eucalyptus globulus* in commercial plantations. IAWA J 22:255–265

Quilhó T, Pereira H, Richter HG (1999) Variability of bark structure in plantation-grown Eucalyptus globulus. IAWA J 20:171–180

Quilhó T, Pereira H, Richter HG (2000) Within-tree variation in phloem cell dimensions and proportions in Eucalyptus globulus. IAWA J 22:255–265

Ragauskas AJ et al (2014) Lignin valorization: improving lignin processing in the biorefinery. Science 344:1246843

Rayonier Advanced Materials (2018) http://rayonieram.com/. Accessed 08 Feb 2018

Reeve DW (2002) The kraft recovery cycle. Tappi kraft recovery operations short course. Tappi Press, Atlanta, U.S.A.

Ritchie TJ, Ertl P, Lewis R (2011) The graphical representation of ADME-related molecule properties for medicinal chemists. Drug Discov Today 16:65–72

Rowell RM (ed) (2005) Handbook of wood chemistry and wood composites. CRC Press, Florida

Saito T, Perkins JH, Vautard F, Meyer HM, Messman JM, Tolnai B, Naskar AK (2014) Methanol fractionation of softwood Kraft lignin: impact on the lignin properties. ChemSusChem 7:221–228

Sannigrahi P, Ragauskas AJ, Miller SJ (2009) Lignin structural modifications resulting from ethanol organosolv treatment of *Loblolly Pine*. Energy Fuel 24:683–689

Santos GG, Alves JCN, Rodilla JML, Duarte AP, Lithgow AM, Urones JG (1997) Terpenoids and other constituents of *Eucalyptus globulus*. Phytochemistry 44:1309–1312

Santos OS, Freire C, Domingues MR, Silvestre A, Neto P (2011) Characterization of phenolic components in polar extracts of *Eucalyptus globulus labill*. Bark by high-performance liquid chromatography–mass spectrometry. J Agric Food Chem 59:9386–9393

Santos SAO, Pinto RJB, Rocha SM, Marques PAAP, Neto CP, Silvestre AJD, Freire CSR (2014) Unveiling the chemistry behind the green synthesis of metal nanoparticles. ChemSusChem 7:2704–2711

Santos SAO, Vilela C, Freire CSR, Neto CP, Silvestre AJD (2013) Ultra-high performance liquid chromatography coupled to mass spectrometry applied to the identification of valuable phenolic compounds from Eucalyptus wood. J Chromatogr B 938:65–74

Santos SAO, Villaverde JJ, Freire CSR, Domingues MRM, Neto CP, Silvestre AJD (2012) Phenolic composition and antioxidant activity of Eucalyptus grandis, E. Urograndis (E. grandis × E. Urophylla) and *E. maidenii* bark extracts. Ind Crop Prod 39:120–127

Sevastyanova O, Lange H, Crestini C, Dobele G, Helander M, Chang L (2014) Selective isolation of technical lignin from the industrial side-streams – structure and properties. In: NWBC – Nordic wood biorefinery conference, Stockholm, 25–27 March

Shatalov AA, Evtuguin DV, Pascoal Neto C (1999) (2-O-α-D-Galactopyranosyl-4-O-methyl-α-D-glucurono)-D-xylan from Eucalyptus globulus Labill. Carbohydr Res 320:93–99

Silva ARG, Errico M, Rong BG (2017) Process alternatives for bioethanol production from organosolv pretreatment using lignocellulosic biomass. Chem Eng Trans 57:1–6

Sjostrom E (1981) Pulping chemistry. In: Wood chemistry: fundamentals and applications. Academic Press, New York, pp 104–145

Sjöström E (1993) Wood chemistry: fundamentals and applications. Academic Press, Inc., London

Smolarski N (2012) High-value opportunities for lignin: unlocking its potential. Frost & Sullivan. https://www.greenmaterials.fr/wp-content/uploads/2013/01/high-value-opportunities-for-lignin-unlocking-its-potential-market-insights.pdf

Snelders J et al (2014) Biorefining of wheat straw using an acetic and formic acid based organosolv fractionation process. Bioresour Technol 156:275–282

Stoklosa RJ, Velez J, Kelkar S, Saffron CM, Thies MC, Hodge DB (2013) Correlating lignin structural features to phase partitioning behavior in a novel aqueous fractionation of softwood Kraft black liquor. Green Chem 15:2904–2912

Stora Enso (2015) http://biomaterials.storaenso.com/AboutUs-Site/Pages/Innovation-centre-for-biomaterials.aspx. Accessed 09 Feb 2018

Stora Enso (2018) Lignin solutions. http://biomaterials.storaenso.com/ProductsServices-Site/Pages/Lignin.aspx. Accessed 09 Feb 2018

Strassberger Z, Tanase S, Rothenberg G (2014) The pros and cons of lignin valorisation in an integrated biorefinery. RSC Adv 4:25310–25318

Sun XF, Jing Z, Fowler P, Wu Y, Rajaratnam M (2011) Structural characterization and isolation of lignin and hemicelluloses from barley straw. Ind Crop Prod 33:588–598

Takuo O (2005) Systematics and health effects of chemically distinct tannins in medicinal plants. Phytochemistry 66:2012–2031

Tanger P, Field JL, Jahn CE, DeFoort MW, Leach JE (2013) Biomass for thermochemical conversion: targets and challenges. Front Plant Sci 4:218

Tao J et al (2016) Effects of organosolv fractionation time on thermal and chemical properties of lignins. RSC Adv 6:79228–79235

Téguia CD, Albers R, Stuart PR (2017) Analysis of economically viable lignin-based biorefinery strategies implemented within a kraft pulp mill. TAPPI J 16:157–169

Teixeira MA, Rodríguez O, Rodrigues AE (2010) Perfumery radar: a predictive tool for perfume family classification. Ind Eng Chem Res 49:11764–11777

Tembec (2002) Lignosulfonates from Tembec. http://kemtekindustries.com/Tembec-Lignosulfonates-Brochure.pdf. Accessed 08 Feb 2018

Valdivia M, Galan JL, Laffarga J, Ramos J-L (2016) Biofuels 2020: biorefineries based on lignocellulosic materials. Microb Biotechnol 9:585–594

Vankar PS, Srivastava J, Molčanov K, Kojic-Prodić B (2009) Withanolide a series steroidal lactones from Eucalyptus globulus bark. Phytochem Lett 2:67–71

Vázquez G, Fontenla E, Santos J, Freire MS, González-Álvarez J, Antorrena G (2008) Antioxidant activity and phenolic content of chestnut (*Castanea sativa*) shell and Eucalyptus (*Eucalyptus globulus*) bark extracts. Ind Crop Prod 28:279–285

Vázquez G, González-Alvarez J, Santos J, Freire MS, Antorrena G (2009) Evaluation of potential applications for chestnut (*Castanea sativa*) Shell and Eucalyptus (*Eucalyptus globulus*) bark extracts. Ind Crop Prod 29:364–370

Vázquez G, Santos J, Freire M, Antorrena G, González-Álvarez J (2012) Extraction of antioxidants from Eucalyptus (*Eucalyptus globulus*) bark. Wood Sci Technol 46:443–457

Verma M, Godboutl S, Brar SK, Solomatnikova O, Lemay SP, Larouche JP (2012) Biofuels production from biomass by thermochemical conversion technologies. Int J Chem Eng 2012:1–18

Wallmo H (2017) The next generation LignoBoost – tailor made lignin production for different lignin bio-product market. In: 7th Nordic wood biorefinery conference (NWBC), Stockholm, Sweden, March 28–30

Wen JL, Xue BL, Sun SL, Sun RC (2013a) Quantitative structural characterization and thermal properties of birch lignins after auto-catalyzed organosolv pretreatment and enzymatic hydrolysis. J Chem Technol Biotechnol 88:1663–1671

Wen JL, Sun SL, Xue BL, Sun RC (2013b) Recent advances in characterization of lignin polymer by solution-state nuclear magnetic resonance (NMR) methodology. Materials 6:359–391

Wen JL, Xue BL, Xu F, Sun RC (2012) Unveiling the structural heterogeneity of bamboo lignin by in situ HSQC NMR technique. Bioenergy Res 5:886–903

Whitmore TC (1962) Studies in systematic bark morphology. New Phytol 61:191–207

Xiao B, Sun XF, Sun R (2001) Chemical, structural, and thermal characterizations of alkali-soluble lignins and hemicelluloses, and cellulose from maize stems, rye straw, and rice straw. Polym Degrad Stab 74:307–319

Xu F, Sun JX, Sun R, Fowler P, Baird MS (2006) Comparative study of organosolv lignins from wheat straw. Ind Crop Prod 23:180–193

Zhao L-Q, Sun Z-H, Zheng P, He J-Y (2006) Biotransformation of isoeugenol to vanillin by *Bacillus fusiformis* CGMCC1347 with the addition of resin HD-8. Process Biochem 41:1673–1676

Chapter 2
Integrated Process for Vanillin and Syringaldehyde Production from Kraft Lignin

Abstract Lignin is one of the main components of pulping liquors; its heterogeneous molecular structure constitutes a valuable source of chemicals, particularly phenolics. However, lignin depolymerization with selective bond cleavage is the major challenge for converting it into value-added chemicals and to accomplish its subsequent valorization. One of the routes for the potential evaluation of lignin as source of syringaldehyde (Sy) and vanillin (V) is toward environmentally friendly processes as oxidation with O_2. In this chapter, the performance of oxidative depolymerization of lignins and black liquors from different sources is evaluated. The effect of reaction conditions on the product yield and the study of kinetic laws were also detailed.

Membrane separation and chromatographic processes suggested as downstream processes to treat the oxidized lignin media are also briefly described. Membrane separation studies refer to studies performed with synthetic lignin/vanillin solutions and a real oxidized lignin solution and address the rejection coefficients toward lignin and phenolate monomers of interest, main fouling factors, and cleaning. Tubular ceramic membranes with molecular weight cutoffs ranging from 50 to 1 kDa were used. The chromatographic processes encompass studies performed with ionic and nonionic resins employing either synthetic mixtures or real oxidized lignin solutions.

Finally, the integrated oxidation and separation process proposed aiming the complete valorization of lignin in phenolic monomers and oligomers is detailed.

Keywords Oxidation of lignin · Batch oxidation · Structured packed bed · Membrane separation · Ion exchange · Adsorption and desorption · Vanillin · Syringaldehyde · Kraft black liquor · Modeling

© Springer Nature Switzerland AG 2018
A. E. Rodrigues et al., *An Integrated Approach for Added-Value Products from Lignocellulosic Biorefineries*, https://doi.org/10.1007/978-3-319-99313-3_2

2.1 Oxidation of Lignin with O₂ in Alkaline Medium

Pulp and paper industries generate large amounts of lignin-based material that are mainly handled in a destructive way to energy production. However, several favorable economic and environmental aspects point out the use of lignocellulosic biomass as an interesting source for obtaining valuable fine chemicals by means of controlled lignin depolymerization. The type and yields of products from lignin depolymerization are highly dependent on the selected process, reaction conditions, and the nature of the raw material. The nature of raw material has great influence in the type and yields of products since the ratio between guaiacyl (G), syringyl (S), and p-hydroxyphenyl (H) units, the molecular weight, and the amount of lignin differ among groups of plants. Softwood lignins primarily contain G units and small proportions of H units, and hardwood lignins contain both S and G units, with a minor proportion of H units. Moreover, even in the same group of plants, in this case hardwoods, there is a high variety of proportions between G and S units as detailed in the literature (Santos et al. 2011).

Literature is extensive regarding the conversion of lignin from different origins into functionalized phenolic monomers of great interest. Lignin depolymerization with selective bond cleavage is the major challenge for converting lignin into value-added chemicals and to accomplish its subsequent valorization. A variety of depolymerization methods have been proposed, and some reviews about this subject have been published (Mahmood et al. 2016; Pandey and Kim 2011; Schutyser et al. 2018). Pyrolysis (thermolysis), gasification, hydrogenolysis, chemical oxidation, and hydrolysis under supercritical conditions are the major thermochemical methods studied (Azadi et al. 2013; Pandey and Kim 2011).

The oxidative depolymerization of lignin can be performed through different types of oxidants, and the characteristics of each oxidant determine their activity and selectivity in the reaction (Lin and Dence 1992). Consequently, the oxidant selection is based on the properties that allow obtaining the maximum product yields from lignin oxidation. Between them it is possible to find oligomeric products, phenolic and non-phenolic compounds. However, in most studies, the authors focus their attention on the identification and quantification of specific phenolic compounds of greatest interest that usually include several aldehydes (V and Sy, Fig. 2.1), acids (vanillic and syringic acid), and ketones (acetovanillone and acetosyringone).

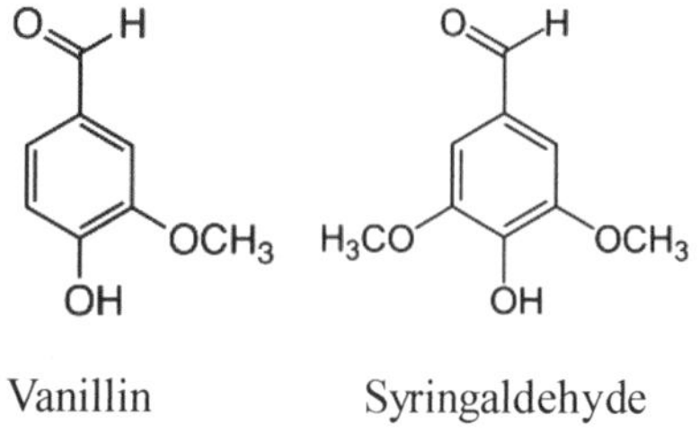

Fig. 2.1 Structure of Sy (3,5-dimethoxy-4-hydroxybenzaldehyde) and V (3-methoxy-4-hydroxybenzaldehyde)

Vanillin   Syringaldehyde

The high yields reported, considering mainly V and Sy, were obtained by oxidation with molecular oxygen in alkaline medium (Pandey and Kim 2011; Pinto et al. 2012). The use of molecular oxygen as oxidant is advantageous when economic and environmental questions are considered. This is an inexpensive and green oxidant, which preserves the lignin aromatic rings during the oxidation reaction. However, when oxygen is used, it is important to take into account some limitations: its non-selectivity, the possibility of over oxidation, its low solubility in the reaction medium, and the requirement of high temperatures for activation (Marshall and Sankey 1951; Pinto et al. 2012; Tromans 1998). Moreover, it is well-known that the phenolic units and ring-conjugated structures of lignin became more reactive with oxygen.

2.1.1 Batch Oxidation

In the last decades, several authors have been working on lignin batch oxidation in order to develop an effective process of V and Sy production from different sources of lignin (Araújo et al. 2010; Mathias et al. 1995a; Mathias and Rodrigues 1995; Pinto et al. 2013; Sales et al. 2006; Santos et al. 2011; Tarabanko et al. 2001). Alkaline oxidation of lignin using O_2 with or without the presence of a catalyst is the most studied process.

The oxidative depolymerization of lignin involves the cleavage of aromatic rings, aryl ether linkages, and/or other linkages in lignin structure. Products from lignin alkaline oxidation are predominantly aromatic aldehydes or acids depending on the reaction conditions. Among them, V and Sy and other phenolics like p-hydroxybenzaldehyde, vanillic, and syringic acids are the most important, and their occurrence plays a decisive role to determine the reaction efficiency (Pinto et al. 2012; Wu et al. 1994). In the literature, it is possible to find different mechanisms that intend to explain the aldehyde formation from lignin precursors by alkaline oxidation. A study about the improvement of the process conditions of oxidative cleavage of an aspen wood lignin into aromatic aldehydes (V and Sy) was developed by Tarabanko and Petukhov (Tarabanko and Petukhov 2003; Tarabanko et al. 2004). In this work, a mechanism for the formation of aromatic aldehydes from lignin oxidative depolymerization was described (Fig. 2.2), which starts from the formation of phenoxyl radical and is accomplished by the formation of V through the cleavage of substituted coniferyl aldehyde. The proposed mechanism allows inferring that the selectivity of the oxidation process could be achieved using more severe reaction conditions, in order to increase the yield of aromatic aldehydes from lignin (Tarabanko and Petukhov 2003).

A literature overview of some representative alkaline oxidations of lignin using O_2 with or without the presence of catalyst shows that the yields do not exceed 12% w/w$_{lignin}$ and 20% w/w$_{lignin}$ for V and Sy, respectively (Fargues et al. 1996b; Mathias and Rodrigues 1995; Pinto et al. 2011; Pinto et al. 2012; Pinto et al. 2013; Santos et al. 2011; Tarabanko and Petukhov 2003; Wu et al. 1994). However, the lignin

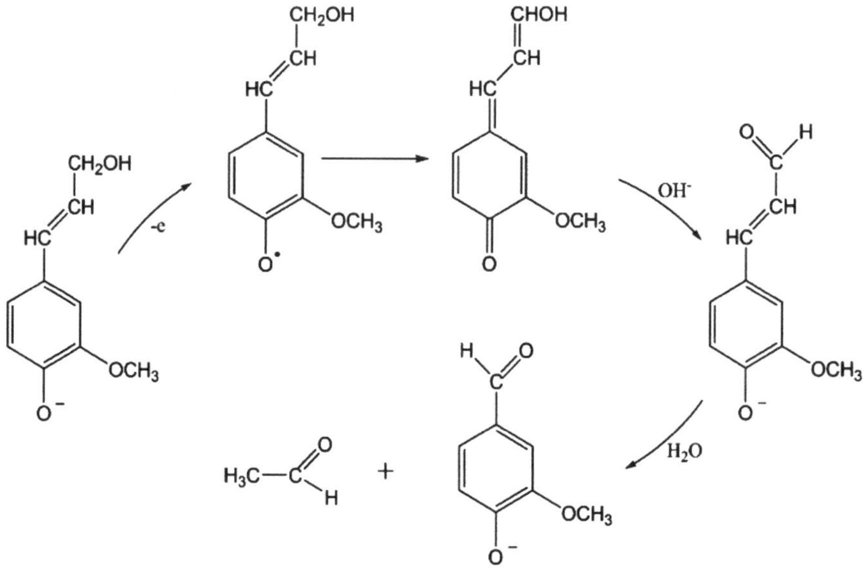

Fig. 2.2 Reaction mechanism for vanillin formation during alkaline oxidation of lignin. (Reprinted by permission from Springer Customer Service Centre GmbH: Springer, Tarabanko et al. (2004), Copyright (2004))

potential for value-added chemical production depends on the experimental conditions, and consequently, several authors studied the effect of lignin oxidation reaction conditions on the production of V and Sy (Araújo et al. 2010; Costa et al. 2015; Pinto et al. 2013; Santos et al. 2011). They have concluded that the highest yields of these phenolic compounds are obtained at high pH (almost 14), high temperatures (higher than 373 K), and using molecular oxygen with a partial pressure equal or higher than 3 bar. The limitation of this process is the low solubility of oxygen in the reaction medium of sodium hydroxide (NaOH) and lignin in the high operational temperatures (Pinto et al. 2012).

Fargues and co-workers studied the production of V from kraft *Pinus spp.* lignin oxidation and found a maximum yield of V when oxidation reaction was performed with lignin concentration of 60 g/l and temperature of 393 K, in alkaline medium containing NaOH at 80 g/l under pO_2 of 4 bar. (Fargues et al. 1996a, b). Moreover, in a study about the oxidation of hardwood kraft lignin to phenolic derivatives with O_2, Villar and co-workers stated that in alkaline oxidation of lignin, the sample has an important effect on the phenolic products yield, as well as the method used for the isolation of lignin (Villar et al. 2001). The authors attributed the low aldehyde yield obtained to the condensed nature of lignins and also to the lignin and lignin oxidation product transformation into low molecular weight acids; therefore, lignin extracts that maintain a structure similar to the native lignin have higher yields (Villar et al. 2001).

What concerns the presence of catalysts in lignin oxidation with O_2 in alkaline medium, the most frequently used are transition metal salts as CuO, $CuSO_4$, $FeCl_3$, and Fe_2O_3, with $CuSO_4$ being the most frequently used (Pinto et al. 2012; Schutyser et al. 2018). The referred catalysts have high oxidation potential and allow electron transference from the aromatic rings of lignin; at the same time, this high oxidation potential makes the regeneration of the metal salt in the catalytic cycle difficult (Pinto et al. 2012). In the literature, the lignin oxidation with catalyst has been studied mainly on model compounds, most of them monomers. Some relevant works about the effectiveness of catalysts in lignin oxidation reported that the addition of a catalyst not only accelerates the reaction but in general also increases the aldehydes yield (Behling et al. 2016; Ma et al. 2015; Sippola and Krause 2005; Wu and Heitz 1995; Zakzeski et al. 2010; Zhang et al. 2009). However, some studies have observed no yield increase in catalyzed oxidations (Mathias and Rodrigues 1995; Villar et al. 2001).

The great interest in lignin chemical oxidation leads to a constant exploration of process modifications and improvements, bringing new highlights in unexplored aspects of the reaction engineering and products (Mathias and Rodrigues 1995; Pinto et al. 2011; Santos et al. 2011; Tarabanko and Petukhov 2003).

2.1.1.1 Kinetics and Modeling of Reaction in Batch Reactor for Vanillin Production

Since the goal of alkaline oxidation is to achieve the maximum conversion into phenolic compounds, several studies are focused in the discussion of the effect of reaction conditions in product yields obtained from lignin and/or spent liquor oxidation (Araújo et al. 2010; Dardelet et al. 1985; Fargues et al. 1996a; Mathias and Rodrigues 1995). Moreover, the accurate selection of the oxidation conditions is important not only to achieve the maximum yields products but also to avoid the oxidation of the produced aldehydes into organic acids such as formic, acetic, lactic, oxalic, syringic, vanillic, and p-hydroxybenzoic. In the literature, the dependency of the phenolic compound yields on temperature, reaction time, O_2 partial pressure (pO_2), initial lignin concentration, and alkaline condition was often studied (Araújo 2008; Fargues et al. 1996a; Mathias 1993; Pinto et al. 2012; Sales et al. 2006; Santos et al. 2011). In general, it was observed that all the parameters have a positive effect on the net conversion of lignin as well as the product yields, except for the reaction time which should have an optimum for the production of the desired compound.

Fargues and co-workers studied the process optimization of the production of V from the oxidation of a kraft *Pinus spp.* lignin (Fargues et al. 1996a, b). The kinetic study was carried out to measure reaction orders with respect to lignin, oxygen, and alkalinity, as well as the influence of temperature on the kinetic rate constants. They proved the dependency of the kinetic constant of V production with the temperature and also show that V produced is also degraded by oxidation whose importance depends on the pH and the temperature of the solution. On the other hand, Araújo and co-workers (Araújo et al. 2010) observed that whatever the lignin source, the

Table 2.1 Isothermal model for lignin oxidation and vanillin production in batch reactor (Mathias 1993; Sridhar et al. 2005)

Rate equation for vanillin production	Rate equation for vanillin oxidation
$r_1 = k_1\,[C_{O2}]^{1.75}[C_{Li}]$	$r_2 = k_2\,[C_{O2}]\,[C_V]\ (\text{pH} > 11.5)$
$k_1 = 1.376 \times 10^7 \exp\left(-\dfrac{3520}{T}\right)$	$k_2 = 4.356 \times 10^6 \exp\left(-\dfrac{5530}{T}\right)$

Overall rate of vanillin production

$r_3 = k_1\,[C_{O2}]^{1.75}[C_{Li}] - k_2\,[C_{O2}]\,[C_V]$

Dissolved oxygen concentration correlation

$$[O_2] = \left(3.559 - 6.659 \times 10^{-3}\,T - 5.606 P_{o2} + 1.594 \times 10^{-5}\,P_{o2}T^2 + 1.498 \times 10^3\,\frac{P_{o2}}{T}\right)\left(10^{-0.144\,t}\right)\left(10^{-3}\right) \text{mol}\,/\,\text{L}$$

Lignin mass balance	Vanillin mass balance
$\dfrac{dC_{Li}}{dt} = -k_1\left[C_{O2}\right]^{1.75}\left[C_{Li}\right]$	$\dfrac{dC_v}{dt} = -k_1\left[C_{O2}\right]^{1.75}\left[C_{Li}\right] - k_2\left[C_{O2}\right]\left[C_V\right]$
Kinetic constant of vanillin production	**Kinetic constant of vanillin oxidation**
$k_1 = 1.376 \times 10^7 \exp\left(-\dfrac{3502}{T}\right)$	$k_2 = 4.356 \times 10^6 \exp\left(-\dfrac{5530}{T}\right)$
$(\text{L/mol})^{1.75}/\text{min}$	L/(mol. min)

Kinetic constant of vanillin oxidation (considering the oxidation of vanillin alone)

$k_2 = 3.61 \times 10^6 \exp\left(-\dfrac{9706}{T}\right) \text{L}\,/\left(\text{mol.min}\right)$

C_{Li} lignin concentration (mol/l), C_{O2} oxygen concentration (mol/l), C_V vanillin concentration (mol/l), k_1 kinetic rate constant (l/mol$^{1.75}$/min), k_2 kinetic rate constant (l/mol/min), P_{O2} partial pressure of O_2 (atm), r_1 rate of vanillin production (mol/l/min), r_2 rate of vanillin oxidation (mol/l/min) r_3 rate of overall vanillin production (mol/l/min), T temperature (K), t time (s)

variation of pH is the most important condition in the oxidation reaction: lower values of pH increase the rate of V degradation, reducing its yield. The kinetic model, developed by these authors, to describe V production in batch reactor is shown in Table 2.1 and was based on the following assumptions:

(a) The experimental kinetic study was entirely based on the concentration of vanillin produced as a function of time, as it was not possible to measure the concentration of lignin nor other products of the oxidation.
(b) The concentration of dissolved oxygen in the solution was calculated from a semiempirical correlation.
(c) The lignin is composed of precursors that will create only vanillin.
(d) The reaction of lignin oxidation into vanillin is irreversible.
(e) The production rate of the vanillin is the same for all the fragments of the lignin.
(f) The reactor is perfectly mixed.
(g) There is no mass transfer limitation from the gas to liquid phase.
(h) Isothermal operation.

Moreover, the mathematical model depends on the lignin source, and the temperature and O_2 pressure should be adjusted for the different reactions taking into account that the final yields are the balance between improving the V conversion and minimizing its oxidation.

Araújo also studied lignin oxidation in batch model and developed a more complex model considering the energy balance of the system (Table 2.2) (Araújo 2008). The assumptions made for this non-isothermal model are (a) perfectly mixed reac-

Table 2.2 Non-isothermal model for lignin oxidation and vanillin production in batch reactor (Araújo 2008; Araújo et al. 2010)

Mass balance equation of the components

Vanillin: $\dfrac{dn_V}{dt} = (r_1 - r_2)V_l - \dfrac{n_v}{V_l}\dfrac{\Delta V_l}{\Delta t}$ / **Lignin:** $\dfrac{dn_L}{dt} = \alpha r_l V_l - \dfrac{n_L}{V_l}\dfrac{\Delta V_l}{\Delta t}$

Sodium hydroxide: $\dfrac{dn_{NaOH}}{dt} = -\dfrac{n_{NaOH}}{V_l}\dfrac{\Delta V_l}{\Delta t}$ / **Water:** $\dfrac{dn_{H_2O}}{dt} = -\dfrac{n_{H_2O}}{V_l}\dfrac{\Delta V_l}{\Delta t}$

Gases in the gas phase (oxygen, nitrogen and water vapor)

$P_{O_2} = P - P_{H_2O} - P_{N_2}$ / $n^g_{O_2} = \dfrac{P_{O_2}V_g}{RT}$ / $P_{N_2} = \dfrac{n_{N_2}RT}{V_g}$ / $P_{H_2O} = \exp\left(A_0 + \dfrac{B_0}{C_0 + T}\right)$

Dissolved oxygen in the interface

$C^*_{O_2} = \left(3.559 - 6.659 \times 10^{-3}T - 5.606 P_{O_2} + 1.594 \times 10^{-5} P_{O_2}T^2 + 1.498 \times 10^3 \dfrac{P_{O_2}}{T}\right)10^{-3} \times 10^{-0.1441}$

Energy balance to the system

$Q - r_1 V_l \Delta H_{r,1} - r_2 V_l \Delta H_{r,2} = \lambda^{H_2O}_{vap}\left(\left[\dfrac{-B_0}{(C_0 + T)^2} - \dfrac{1}{T}\right]\dfrac{V_g P_{H_2O}}{RT}\dfrac{dT}{dt} + \dfrac{\Delta V_l}{\Delta t}\dfrac{P_{H_2O}}{RT}\right) + \Sigma n_i Cp_i \dfrac{dT}{dt}$

Heat exchanged between the system and the surroundings

$Q = U_1 A_1 (T_F - T) - U_2 A_2 (T - T_{amb})$

A_1 internal area of the cylindrical glass limiting the reaction zone (m²), A_2 surface area of the steel plates (m²), A_0 B_0, C_0 parameters of the Antoine equation relating vapor pressure to temperature, $C_{O_2}^*$ concentration of oxygen in the liquid in the gas-liquid interface (mol/m³), Cp_i heat capacity of substance i (J kg⁻¹ K⁻¹), n_{H_2O} number of moles of water (mol), n_L number of moles of lignin (mol), n_{NaOH} number of moles of sodium hydroxide (mol), n_V number of moles of vanillin (mol), P total pressure (bar), P_{H_2O} partial pressure of water (bar), P_{N_2} partial pressure of nitrogen (bar), P_{O_2} partial pressure of oxygen (bar), Q heat received by the system from the surroundings (W), R universal gas constant (l atm mol⁻¹ K⁻¹), r_1 rate of formation of vanillin (mol m⁻³ s⁻¹ or mol m⁻³ min⁻¹), r_2 rate of oxidation of vanillin (mol m⁻³ s⁻¹ or mol m⁻³ min⁻¹), T reaction temperature (K), t time (s), T_F thermofluid temperature inside the jacket (K), $U1$ overall heat transfer coefficient from the thermofluid in the jacket to the liquid inside the reactor (Wm⁻² K⁻¹), U_2 overall heat transfer coefficient from inside of the reactor to the ambient through the top and bottom stainless steel plates (Wm⁻² K⁻¹), V_g volume of the gas phase inside the reactor (m³), V_l volume of the liquid phase inside the reactor (m³), α lignin stoichiometric coefficient on the lignin oxidation reaction, $\Delta H_{r,1}$ heat of reaction of lignin oxidation (J/mol), $\Delta H_{r,2}$ heat of reaction of vanillin oxidation (J/mol), Δt time interval between collection of two consecutive liquid samples (s), ΔV_l volume of liquid taken from the system in each sample collection (m³), $\lambda^{H_2O}_{vap}$ heat of vaporization of water (J/mol)

tor: constant spatial values of all variables within the reactor system; (b) constant total pressure during the reaction; (c) mass transfer resistances are negligible; (d) irreversible oxidation reactions; and (e) ideal behavior of the gas phase.

Santos and co-workers (Santos et al. 2011) studied the major products obtained from the oxidation of an eucalyptus lignosulfonate (LS) and the kinetics of their formation. The oxidation reactions were performed with molecular O_2, in alkaline medium, and in the temperature range of 403–423 K. The authors observed that oxidation of LS leads to a predominant formation of Sy and V among low molecular weight aromatic oxidation products. The kinetic results proved that the rate constant of Sy formation was more than twice that for V and that Sy also suffered faster degradation (about 5 times) than V, under the same conditions. In this study, the authors also suggested that aromatic aldehydes in LS oxidation, under alkaline conditions, are formed via different mechanisms than aromatic acids and that their yields are drastically affected by carbohydrates, which should be eliminated from sulfite spent liquor before oxidation (Santos et al. 2011). Moreover, Fargues and co-workers (Fargues et al. 1996a) referred that, in lignin oxidation reactions, pO_2 should not be too high and the effect of pO_2 is on the rate of product formation with no influence on the yields. Nevertheless, the pO_2 should be controlled to avoid further oxidation of V (Collis 1954; Fargues et al. 1996b). Changes in total pressure from 5 to 14 bar do not affect the products yields (Fargues et al. 1996a).

Some authors also referred that the reaction time and temperature should also be controlled to avoid degradation of the aldehydes produced leading to the formation of acids (Fargues et al. 1996a; Mathias 1993). On the other hand, it is also stated in the literature that higher temperatures allow obtaining higher V yields in a shorter reaction time. Considering that V yield has a maximum with regard to the reaction severity, the rate of degradation of this compound is also higher (Pinto et al. 2012; Santos et al. 2011). Sales and co-workers studied lignin degradation reactions and aromatic aldehydes formation using a kinetic model quantified by a complex reaction network (Sales et al. 2006). Under the selected reaction conditions of temperature (in the range of 373–413 K) and pO2 (between 2 and 10 bar), the kinetic evolution of a sugarcane lignin oxidation and product formation was investigated. The authors demonstrated that moderate pO2 and short reaction times must be employed in order to obtain the maximum yields of intermediate oxidation products, such as V, Sy, and Hy, and to avoid that the produced aldehydes are oxidized into organic acids, since the lignin consumption is a faster reaction step.

Another important process parameter for aldehyde production is the pH value of the mixture. pH should stay higher than 12 in order to keep the high alkalinity during the reaction and consequently for the total ionization of phenolic groups and conversion to reactive quinone methide. During the lignin oxidation process, the yield of V decreases when the pH value begins to decrease. Indeed, high alkali concentrations (pH >12) reduce V degradation, whereas at lower pH values (<11.5), an accentuated decrease in V yield is observed (Fargues et al. 1996a; Pinto et al. 2012). This phenomenon was attributed to the protonation of reaction intermediates, more basic than the phenolics produced (vanillin pKa = 7.4) (Fargues et al. 1996a; Pinto et al. 2012; Tarabanko and Petukhov 2003).

2.1.1.2 Syringaldehyde as the Main Product from Hardwood Lignins

The demand for chemicals from renewable sources has brought an increasing interest on lignin route to produce Sy (3,5-dimethoxy-4-hydroxybenzaldehyde). Sy is a promising aromatic aldehyde very similar in structure to V with the advantage of already containing two methoxyl groups (German Federal Government 2012; Holladay et al. 2007). This aldehyde is used as flavor and fragrance ingredient in food and cosmetic industries and also as precursor of second-generation fine chemicals and drugs in pharmaceutical industry (Erofeev et al. 1990; Ibrahim et al. 2012; Manchand et al. 1992) such as 3,4,5-trimethoxybenzaldehyde, a building block of the antibacterial agent ormetoprim and trimethoprim. Sy can be synthesized from gallic acid, pyrogallol, vanillin, or p-cresol (Ibrahim et al. 2012), and its market value is about 22 USD per kilogram (Christopher 2012).

Lignin has become less attractive as a raw material for V production due to the developments in the pulp and paper industry, such as the replacement of softwoods by hardwoods and non-wood plants as desirable raw materials (Hocking 1997; Pinto et al. 2013).

The type of phenolic compounds produced from lignin depolymerization is determined by the precursor content in lignin structure. G lignins (typical of softwoods) yield V under oxidative depolymerization, whereas H:G:S lignins (typical of non-wood plants and hardwoods) are able to produce Sy, Hy, and lower yields of V (Pinto et al. 2011; Sales et al. 2006; Villar et al. 2001; Wong et al. 2010a; Wu et al. 1994). Consequently, lignin oxidation has been studied as a way to produce Sy, being an important approach in view of lignin exploitation for high-value-added applications (Costa et al. 2015; Pinto et al. 2011; Pinto et al. 2013; Santos et al. 2011; Wu et al. 1994). At the same time, this could be seen as a disadvantage since the process (reaction and separation schemes) and market were focused on V. However, the implementation of new practices depends on its economic sustainability which, in turn, depends largely of the raw material availability and product yield (from reaction to final separation).

2.1.1.3 Oxidation of *Eucalyptus globulus* Kraft Pulping Liquor Versus Kraft Lignin

Considering the availability of side streams from pulp industries and biorefineries, it could be interesting to explore the industrial potential of pulping liquors as source of high-value-added chemicals. The potential evaluation of pulping liquors versus lignins would be an important tool for the improvement of biorefining activity that claims for valorization due to environmental and economic factors. However, works focused on the depolymerization of isolated and nonisolated lignins are difficult to find in the literature. Pinto and co-workers (Pinto et al. 2013) developed a complete study to evaluate the potential of *E. globulus* kraft pulping liquors and lignins as source of Sy and V and other minor products. The authors studied three industrial kraft liquors from different processing stages: (1) at the digester outlet (KL), (2) after evaporation (EKL), and (3) after heat treatment (HTEKL) and the respective isolated lignins. The potential of liquors and lignins was evaluated through oxidation with O_2 in alkaline medium. For each lignin and liquor, the profiles of low molecular weight phenolics were established, as well as the yields, selectivity, time to maximum, temperature, and O_2 uptake during the reaction. This study was focused in the yields of Sy, V (the main phenolic products identified in the oxidation mixture), and the respective carboxylic acids, syringic and vanillic acids. The authors observed that the progress of the product yield with reaction time does not significantly differ among the liquors and lignins. However, a lower uptake of O_2 is observed in the direct oxidation of kraft liquors, which is point out as a consequence of its high content on inorganic compounds comparatively to isolated lignins. The high content of inorganic components in liquors reduces the consumption of O_2 during oxidation and consequently limits the product formation, decreasing their yields (Pinto et al. 2013). Moreover, more than the carbohydrate and inorganic contents, in oxidation reactions where the alkali charge and O_2 are not limiting factors, product yield is primarily affected by lignin structure, due to the contribution of condensed structures.

Pinto and co-workers (Pinto et al. 2013) have also demonstrated that KL has lower content on condensed structures and higher potential than the other kraft liquors for aldehydes production; moreover, the isolation of lignin is advantageous only in the case of this liquor, leading to an increase of about 20% and 25% in the yields of Sy and V, respectively. For the others, liquors and lignins (EKL, EKLlig, HTEKL, and HTEKLlig), it was observed that the additional yield accomplished by oxidizing the isolated lignins is not enough to overcome the low recovery yield of lignins during the isolation process, concluding that lignin isolation benefits yields and selectivity but just in the case of KL. In Table 2.3, it is possible to find Sy and V yields per ton of nonvolatile solids for each liquor and lignin; the values confirm that the productivity is rather lower for direct oxidation of evaporated and heat-treated liquors.

In the case of KL, the balance between the cost of isolation and the extra value obtained from the higher yields should be decisive if this route for lignin valorization is envisaged (Pinto et al. 2013). Moreover, it is important to take into consideration that when product yield from oxidation of *E. globulus* kraft pulping liquor and kraft lignins is evaluated, it is essential to understand the benefit of lignin isolation, considering that the pulping liquors could be directly oxidized; consequently, the benefit of prior lignin separation should be carried out for each case taking the yields, selectivity, O_2 consumption, and time as assessment factors.

Table 2.3 Syringaldehyde and vanillin yields per ton of nonvolatile solids (Pinto et al. 2013)

| | Sample | kg of product per ton of solids | |
		Sy	V
Liquors	KL	8.6	3.5
	EKL	6.4	3.3
	HTEKL	6.9	2.5
Lignins	KLlig	10.3	4.4
	EKLlig	4.1	1.6
	HTEKLlig	2.8	1.3

2.1.2 Oxidation in Cocurrent Gas-Liquid Flow Structured Packed Reactor

After the research on lignin oxidation in batch mode, the interest is focused in the continuous process. From an industrial point of view, the continuous process of lignin oxidation represents a huge advantage since a large volume of kraft liquor is generated in a pulp and paper industry. Moreover, continuous processes are easier to control and to attain constant product characteristics, and their overall investments and operating costs are usually lower (Borges da Silva et al. 2009).

An experimental pilot setup was built in LSRE-LCM for lignin oxidation in continuous operating mode (Araújo 2008; Araújo et al. 2009). The authors selected a bubble column reactor based on the reasons as follows: no shaft sealing is required, enabling the operation of aggressive media at high temperatures and pressures, reasonable price and can be easily adapted and resized, and can provide uniform temperature throughout even with strong exothermic reactions (Araújo et al. 2009). Figure 2.3 shows the schematic diagram of the operational pilot installation built. The structured packed bed continuous reactor (SPBCR) has a capacity of about 8 L, and the main body of the reactor was filled with Mellapak 750Y structured packing from Sulzer Chemtech (Switzerland) to improve the overall mass transfer performance of the system. More details on the reactor setup can be found in the literature (Araújo 2008; Borges da Silva et al. 2009).

In a typical continuous process of oxidation, the alkaline solution of lignin (60 g/L of lignin in NaOH 2 M) is fed to SPBCR by a piston pump. The temperature inside the reactor was regulated to 403 K, and N_2 is used to pressurize the system. The oxidation is initiated when the operating temperature, pressure, and flow rates stabilized. According to the results obtained by the authors, it was possible to observe that the conversion of lignin to vanillin in all the oxidation experiments performed was substantially smaller than the conversion obtained in the batch reactor. With the referred lignin oxidation conditions, the continuous operation led to about 25–30% of the maximum of vanillin concentration produced in the batch process. The main reason pointed by the authors for this difference is the poor mass transfer of oxygen from gas to liquid phase. Moreover, in order to improve the reaction rate, it is suggested that a higher mass transfer coefficient is required in order to achieve a high conversion of lignin to vanillin.

2.1.2.1 Experimental and Modeling of Vanillin Production

To better understand the reaction and transport phenomena that occur during lignin oxidation experiments on SPBCR, a mathematical model was developed in LSRE-LCM. The mathematical model developed was employed to describe experimental lignin oxidation data in SPBCR with and without the internals and used to determine the effect of the main operating conditions to enhance the reactor performance (Araújo 2008).

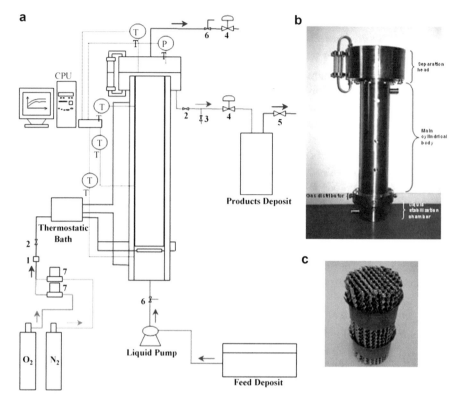

Fig. 2.3 (**a**) Schematic diagram of the pilot setup for the continuous production of vanillin from lignin (1, safety valve; 2, on-off valve; 3, electromagnetic valve; 4, needle valve; 5, safety valve; 6, three-way valve; 7, mass flow controller; PT, pressure transducer; TT, thermocouple); (**b**) Structured bubble column reactor (both the cylindrical main body and the liquid stabilization chamber are jacketed); (**c**) Mellapak 750.Y packing module (made in 316 L stainless steel). (Reprinted from Borges da Silva et al. (2009), Copyright (2009), with permission from Elsevier)

Araujo and co-workers (Araújo 2008; Araújo et al. 2009) defined the mathematical model considering the following assumptions:

1. Constant gas composition with the axial position and time.
2. Isobaric reactor.
3. The oxygen gas-liquid mass transfer is dominated by the resistance in the liquid film.
4. Ideal behavior of the gas phase.
5. For the same axial position, the temperatures of the packing, gas, and liquid phases are equal (pseudo-homogeneous model for the energy balance).
6. The external tube of the reactor jacket is thermally insulated from the surroundings, and the temperature of the outer tube of the column is equal to the temperature of the thermofluid flowing inside the jacket.
7. No heat losses in the thermofluid between the exit of the bath and the entrance of the reactor jacket.

The complete set of equations defining the dynamic nonisothermal model for lignin oxidation in the SPBCR is shown in Table 2.4. The effect of some operating conditions was studied using the mathematical model discussed previously in order to improve the performance of the continuous reactor and try to achieve the product yield obtained in batch mode. The effect of liquid feed flow rate (Q_L), gas feed flow rate (Q_G), set point of the thermostatic bath (T_F^{set}), and the oxygen partial pressure (P_{O2}) was evaluated (Araújo 2008). Figure 2.4 presented the simulation results obtained by the authors for the steady state of vanillin concentration obtained in the SPBCR for the different operating conditions studied.

From Fig. 2.4a, it can be observed that higher liquid flow rates, increasing the set point of the thermostatic bath, lead to an increase in the vanillin yield obtained in the exit stream of the reactor. However, this behavior is the opposite to the one predicted by the Araújo (Araújo 2008) for the lower liquid flow rate in the same point range. This can be a consequence of the increase of liquid residence time that degrades more lignin, and the pH starts to decrease into values where the growing tendency of the difference between the reaction rates with temperature inverts. What concerns the variation of the gas feed flow rate (Fig. 2.4b), the author concluded that higher steady states vanillin concentrations in the exit stream were obtained for higher gas flow rates, due to the linear increase of $k_L a$. For the lower liquid residence times (higher liquid flow rates), the degree of vanillin formation is smaller, since oxygen mass transfer rate is still not good enough. On the other hand, Fig. 2.4c shows that an increase in partial pressure of oxygen results in an increase in vanillin yield. The increase of partial pressure of oxygen results in a higher oxygen solubility, which promotes lignin oxidation. Figure 2.4d reveals an increase in the vanillin production with $k_L a$, with the exception of the results obtained for Q_L of 2 l/h. in this case, it seems that there is a maximum level of vanillin production with $k_L a$., and the productivity starts to decrease by undesired vanillin oxidation. The authors of this work suggest that for a given amount of liquid fed to the reactor, there is an oxidative capacity of the media that should lead to a maximum vanillin yield (Araújo et al. 2009).

Based on the complete study gathered above, the authors establish a set of operating conditions that could improve vanillin yield from SPBCR experiments (Araújo 2008). The selected set of conditions was Q_L of 0.167 l/min, T_F^{set} of 433 K, P_{O2} of 10 bar with pure oxygen, P of 10 bar, Q_G of 40 $SLPM$, and $k_L a$ of 1.5×10^{-2} s^{-1}. The simulation results of the vanillin concentration at the exit of the separation head obtained using the proposed set of operating conditions are shown in Fig. 2.5.

The results shown that the steady state value of vanillin concentration in the exit stream was around 1.8 g/L (3% of mass lignin conversion into vanillin). This result shows that the vanillin concentration obtained from the continuous process was significantly improved reaching about 85% of maximum vanillin concentration obtained in batch reactor.

The effect of lignin concentration, lignin molecular weight, and reaction temperature on the vanillin yield obtained from lignin oxidation in continuous process was also evaluated using the mathematical model developed (Sridhar et al. 2005).

Table 2.4 Mathematical model for the lignin oxidation in the SPBCR (Araújo 2008; Araújo et al. 2009)

Mass balance equation

$$\varepsilon_L D_{ax} \frac{\partial^2 C_i}{\partial z^2} = u_{LS} \frac{\partial C_i}{\partial z} - \varepsilon_L \sum_k v_{i,k} r_k + \varepsilon_L \frac{\partial C_i}{\partial t} - k_L a \left(C_i^* - C_i \right)$$

Dissolved oxygen in the interface

$$C_{O_2}^* = \left(3.559 - 6.659 \times 10^{-3} T - 5.606 P_{O_2} + 1.594 \times 10^{-5} P_{O_2} T^2 + 1.498 \times 10^3 \frac{P_{O_2}}{T} \right) \times 10^{-0.144 I}$$

Energy balance to the reaction media

$$\lambda_{ef} \frac{\partial^2 T}{\partial z^2} = \left(u_{LS} \rho_L C_{P.L} + u_{GS} \rho_G C_{P.G} \right) \frac{\partial T}{\partial z} - \varepsilon_L \left[\left(-\Delta H_{R,1} \right) r_1 + \left(-\Delta H_{R,2} \right) r_2 \right] - \frac{2\pi R_2}{A_R} U \left(T_F - T \right)$$

$$+ \left(\varepsilon_L \rho_L C_{P.L} + \varepsilon_G \rho_G C_{P.G} + \varepsilon_S \rho_S C_{P.S} + \rho_W \frac{A_w}{A_R} C_{P.W} \right) \frac{\partial T}{\partial t}$$

Energy balance to the fluid in the jacket

$$\rho_F u_F C_{P.F} \frac{\partial T_F}{\partial z} + \frac{2R_1}{R_3^2 - R_2^2} U \left(T_F - T \right) + \rho_F C_{P.F} \frac{\partial T_F}{\partial t} = 0$$

Separation head (described by two equal stirred tanks)

$$Q_L C_i \big|_{z=0.8m} = Q_L C_{i,1} - V_{ST} \sum_k v_{i,k} r_k + V_{ST} \frac{dC_{i,1}}{dt}$$

$$Q_L C_{i,1} = Q_L C_{i,2} - V_{ST} \sum_k v_{i,k} r_k + V_{ST} \frac{dC_{i,2}}{dt}$$

A_R internal cross-section area of the column (m^2), A_W difference between the external and the internal cross-section area of the column (m^2), C_i concentration of species i in the liquid phase of the column (mol/m^3), C_i^* concentration of species i in the liquid in the gas-liquid interface (mol/m^3), C_{Pi} heat capacity of substance i (J kg^{-1} K^{-1}), D_{ax} axial dispersion coefficient (m^2/s), ε_G gas hold up, ε_L liquid hold up, ε_S volume fraction of the structured packing, $k_L a$ liquid side mass transfer coefficient (s^{-1}), P_{O2} partial pressure of O$_2$ (atm), Q_L liquid flow rate (l/min), R_1 radius of the internal wall of the reactor column (m), R_2 radius of the external wall of the reactor column (m), R_3 radius of the internal wall of the outer jacket tube (m), r_k rate of reaction k, T reactor temperature (K), t time (s), T_F thermofluid temperature inside the jacket (K), U overall heat transfer coefficient from the thermos fluid in the jacket to the liquid inside the reactor (W m^{-2} K^{-1}), u_{GS} superficial gas velocity (m/s), u_{LS} superficial liquid velocity (m/s), V_{ST} volume of each stirred tank (m^3), $\Delta H_{R,1}$ heat of reaction of lignin oxidation (J/mol), $\Delta H_{R,2}$ heat of reaction of vanillin oxidation (J/mol), λ_{ef} effective thermal dispersion coefficient (W m^{-1} K^{-1}), $v_{i,k}$ stoichiometric coefficient of compound i in the reaction k, ρ_i density of substance i (kg/m^3), Subscripts: F thermofluid, G gas phase, L lignin or liquid phase, O_2 oxygen, S structured packing, W wall

The effect of lignin molecular weight, lignin concentration, and reaction temperature on the yield of vanillin is shown in Fig. 2.6.

Lower molecular weight lignin seems to contain more amounts of precursors which are responsible for higher vanillin yields. It is also noted that the simulation

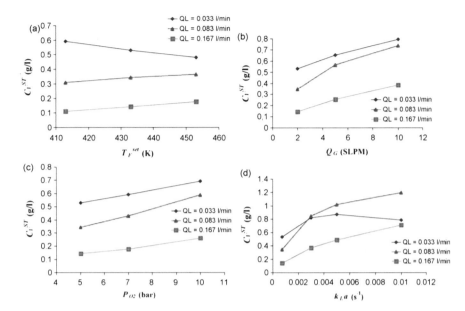

Fig. 2.4 Predicted values of the steady state vanillin concentration in the exit stream (C_V^{ST}) for different T_F^{set} (**a**), Q_G (**b**), P_{O2} (**c**), and k_La (**d**). The fixed values for each case where (**a**) Q_G of 2 SLPM, P_{O2} of 5 bar, and k_La of 7.06×10^{-4} s^{-1}; (**b**) T_F^{set} of 433 K, P_{O2} of 5 bar, and k_La of 7.06×10^{-4} s^{-1} (Q_G of 2 SLPM), 1.36×10^{-3} s^{-1} (Q_G of 5 SLPM), and 2.42×10^{-3} s^{-1} (Q_G of 10 SLPM); (**c**) T_F^{set} of 433 K, Q_G of 2 SLPM, and k_La of 7.06×10^{-4} s^{-1}; (**d**) T_F^{set} of 433 K, Q_G of 2 SLPM, and P_{O2} of 5 bar. The total pressure was 10 bar for all simulations. (Reprinted from Araújo et al. (2009), Copyright (2009), with permission from Elsevier)

Fig. 2.5 Simulation results for the vanillin concentration at the exit stream of the reactor using the selected set of operating conditions. (Reprinted from Araújo et al. (2009), Copyright (2009), with permission from Elsevier)

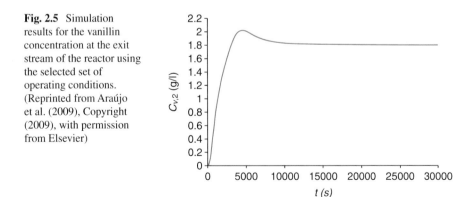

results show higher initial concentrations of lignin lead to higher vanillin yields. The authors found that 60 grams per liter of initial lignin concentration gave an optimum vanillin yield (Sridhar et al. 2005). What concerns the effect of temperature, it was observed that temperature inside the reactor is an important experimental condition that should be considered, since increasing the temperature, all the

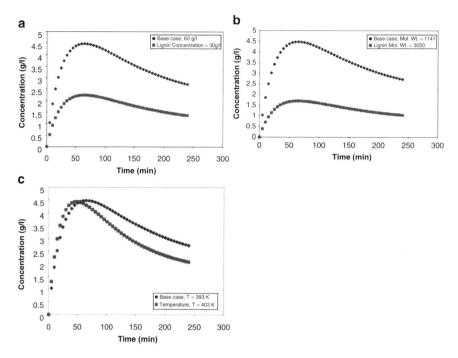

Fig. 2.6 Effect of lignin concentration, effect of lignin molecular weight, and effect of reaction temperature on the yield of vanillin in a SPBCR. (Reprinted from Sridhar et al. (2005), Copyright (2005), with permission from Elsevier)

reaction rates increase, and the solubility of oxygen in the liquid phase is reduced. Consequently, an accurate temperature should accomplish the trade-off between high vanillin conversion and small oxidation of vanillin but maintaining as high as possible the solubility of oxygen in the liquid phase.

The developed mathematical model showed to be a good tool for the selection of the operating parameters for lignin oxidation in SPBCR that enable to obtain the maximum vanillin yields, close to the values obtained in batch mode.

2.1.2.2 Experiments of Oxidation of Hardwood Pulping Liquor and Lignins

There is few information available in the literature about lignin oxidation reaction in SPBCR. However, the research group led by Professor Rodrigues has been developing works in this area for some time.

One of the first experiments of lignin oxidation in continuous process was performed with a kraft lignin Indulin AT supplied by Westvaco Co. The authors (Araújo 2008) evaluated the performance of SPBCR configuration with five lignin oxidation experiments at different liquid and gas flow rates. The operating conditions used in these experiments are summarized in Table 2.5.

Table 2.5 Operating conditions of the lignin oxidations performed in SPBCR (Araújo 2008)

	C_L (g/l)	pH	Q_L (l/h)	Q_{O2}^a (ml$_{NTP}$/min)	Q_{N2}^a (ml$_{NTP}$/min)	P (bar)
exp. 1	60	14	4.20	250	375	10
exp. 2	60	14	4.22	100	150	10
exp. 3	60	14	4.70	1000	1000	10
exp. 4	60	14	2.12	1000	1000	10
exp. 5	60	14	0.99	1000	1000	10

[a]NTP: normal temperature (273 K) and pressure (1 atm). All the experiments were performed with a T_F^{set} of 443 K

For experiments 1 and 2, the steady state was achieved with a vanillin concentration in the exit stream of 0.22 and 0.20 g/l, respectively, which represents a conversion of about 0.37% of the lignin mass into vanillin, a very low value of lignin conversion. Since the range of the gas flow rate used in these two experiments seemed to be insufficient, the authors decided to perform other experiment (exp. 3) with higher gas flow rate. The results showed an increase in vanillin concentration that reaches 0.43 g/l (0.72% of lignin mass conversion into vanillin). However, the yield obtained is still far from the values obtained in the batch reactor, and the authors decided to increase the liquid residence time, by lowering the liquid flow rate (exp. 4). Consequently, the maximum concentration obtained in this experiment was 0.73 g/l, which is a lignin mass conversion into vanillin of 1.22%. What concerns the last experiment (exp. 5) the maximum lignin mass conversion obtained was 1.48%, which corresponds to a vanillin concentration of 0.89 g/l. these values represent an improvement comparatively to the previous experiment, but the liquid flow rate used is smaller. In the end of the study (Araújo 2008), it was concluded that at steady state, reached at approximately 6 h of operation, the highest concentration of vanillin obtained from the SPBCR configuration was 0.89 g/L for exp. 5 that corresponds to 40% vanillin yield from batch oxidation of the same type of lignin (Indulin AT Kraft lignin).

The research group continues to develop work in lignin oxidation in continuous mode with SPBCR using different lignin types and also pulping liquors.

2.2 Separation Processes

2.2.1 Membrane Separation of Phenolates from Depolymerized Lignin

A considerably research effort has been devoted to ultrafiltration of the black liquor mainly focused on lignin concentration or fractionation (Arkell et al. 2014; Evju 1979; Jönsson et al. 2008; Toledano et al. 2010a, b; Uloth and Wearing 1988) in order to obtain more uniform molecular weight lignin fractions and benefit its conversion into chemicals or materials.

Regarding the application of membrane separation to treat oxidized lignin solutions, studies are more recent and restricted to the research studies summarized in Table 2.6. Oxidized lignin solutions are more complex than their respective solutions of origin, containing a wider range of compounds and molecular weight distributions. Depending on the lignin-based raw materials, it is expected that oxidized lignin solutions contain lignin oligomers, simple phenolic compounds (e.g., aldehydes, ketones, and acids), and other secondary products such as lactones, guaiacol, and syringol (Gierer 1986; Pinto et al. 2011; Wong et al. 2010b).

In membrane separation, the oligomers and other higher molecular weight compounds will be retained by the membrane, while the low molecular weight compounds will cross the membrane pores into the permeate stream. In this way, membrane separation studies performed with oxidized lignin streams are important in order to design an adequate industrial process to valorize both retentate and permeate streams generated.

The origin of the lignin and oxidation conditions will result in different solutions that could account for different fouling phenomena during membrane processing, affecting the performance of the separation process in a different way. Moreover, the performance of the separation system is intrinsically related with the chemistry of the membrane surface and solutes once it defines the type of solute-membrane and solute-solute interactions and thus, the type of fouling formed. The former will be responsible for fouling by adsorption of solute on the membrane surface (electrostatic attractions/repulsions), and the latter influences fouling caused by solute aggregation in solution and/or to molecules already adsorbed on the surface of the membrane.

Werhan et al. (2012) have applied nanofiltration (NF) membranes with molecular weight (MW) cutoffs ranging from 0.250 to 0.900 kDa to treat an oxidized Indulin AT reaction medium previously extracted with ethyl acetate. The authors managed to attain different rejections to dimers, trimers, and monomers (Table 2.6). The best separation system was for the 0.600 kDa membrane, the one showing the highest retention to dimers and trimers of 83% and 93%, respectively. However, this membrane also retained the monomers in a higher extent (38%), which accounts for higher mass losses of the desired monomers in the permeate stream.

Regarding the membrane productivities observed, the 0.600 kDa was the one revealing the highest values, and thus the authors have considered that this membrane was the most suitable because of the best overall performance in terms of rejection and flux values observed.

Žabková et al. (2007b) employed tubular ceramic membranes with MW cutoffs of 1, 5, and 15 kDa aiming the vanillate recovery from a synthetic lignin solution prepared with sodium hydroxide to simulate an alkaline oxidized lignin medium. This work was important to demonstrate the viability of treating this type of solutions with ceramic membranes and understand the main phenomena influencing membrane performance.

The selection of the membranes is based on the strong physical chemical resistance of the ceramic membranes, suitable for processing under strong alkaline solutions and high temperatures, such as the solution coming out from the oxidative reactor (pH values between 9–12 and temperature above 100 °C) (Pinto et al. 2011). Moreover, several studies have reported that membranes with an active layer of TiO$_2$/Al$_2$O$_3$/ZrO$_2$, similar to the ones employed by the authors, are suitable membranes to treat alkaline lignin solutions once they have negative charges at high pH values (Almécija et al. 2007; Benfer et al. 2004; de la Rubia et al. 2006) and will favor electrostatic repulsing interactions between the membrane and the ionized solutes. It is expected that the membrane-solute interactions are minimized and will not contribute for flux decline during operation.

In the particular study case of Žabková et al. (2007b), employing lignin in alkaline medium, it is expected to have ionized compounds in solution at a pH of operation (>10) (Dong et al. 1996; Sjöström 1993; Sundin 2000). These solutions contain depolymerized lignin and other low molecular weight compounds in their sodium salt form that are expected to be negatively charged at the operating pH. Dong et al. (1996) study showed that the isoelectric point for kraft lignin is 1.0 and, above this pH, the zeta potential is negative. Additionally, the acid dissociation constants, pKa, of the main low molecular weight compounds present are between 4.3 and 7.9 (Mota et al. 2016c) and thus, it is expected that these compounds have negative charge in strong alkaline solutions. In this study, the authors accomplished lignin retentions superior to 87% and practically no vanillin retentions (Table 2.6). The highest lignin retention coefficients of 97% were accomplished for the 1 and 5 kDa membranes.

The influence of pH was studied and demonstrated a different lignin retention behavior according to the membrane applied. For the 15 kDa membrane, the authors observed that the retention coefficient decreases with the pH increase, while for the 1 kDa, it was observed the opposite. This different rejection behavior can be explained by the different lignin association mechanisms that are highly dependent on the molecular weight, concentration, pH, and ionic strength of the mixture. Moreover, it was also observed that permeate fluxes declined with pH increase explained by the hydrophobicity of membranes and solute molecules. This effect was more pronounced for higher lignin concentration mixtures (Žabková et al. 2007b).

An unsteady-state model was applied to explain flux evolution with processing time considering the osmotic pressure and growth of deposited layer at the membrane surface wall. The mathematical model applied confirmed that the growth of the rejected solute (lignin) on the membrane surface resulted from a progressive increase on the total resistance; the membrane wall concentration was directly associated with bulk concentration, and the estimated wall lignin concentrations suggested the formation of a gel layer for processing with the 5 and 15 kDa membranes.

The initial permeate flux was easily recovered after cleaning with 0.1 M NaOH solution, demonstrating that irreversible fouling is not predominant and membranes can be reused.

In a recent study, the membrane separation of an oxidized industrial kraft liquor from *Eucalyptus* was addressed (Mota 2017), continuing the previous work performed with model solutions (Žabková et al. 2007b). A three-stage membrane fractionation sequence was employed with ceramic ultrafiltration (UF) membranes of MW cutoffs 50, 5, and 1 kDa. The different permeate and retentate streams were characterized

Table 2.6 Membrane separation processes applied aiming the recovery of phenolic monomers from lignin oxidized medium (Mota 2017; Werhan et al. 2012; Žabková et al. 2007b)

Type of solution	Membrane composition	MW cutoff (kDa)	Operational conditions	Rejections (%)
Ethyl acetate extract of oxidized Indulin AT	Cross-linked polyimide	0.280	C_{feed} 5 g/L of lignin oxidation products; 30 °C, 20 bar	8% monomers; 77% dimers; 87% trimers
	Cross-linked polyimide	0.500		16% monomers; 78% dimers; 89% trimers
	PDMS on PAN	0.600		38% monomers; 83% dimers; 93% trimers
	Cross-linked polyimide	0.900		23% monomers; 81% dimers; 92% trimers
Synthetic mixture of vanillin and commercial lignin with 60 k g/mol	TiO_2	1	1.2 m/s; C_{feed} 60/6 g L^{-1} lignin/vanillin; pH 12.5; 25 °C; 1.55 bar;	Lignin (97%); vanillin ($\approx$0)
			1.3 m/s; C_{feed} 60/6 g L^{-1} lignin/vanillin; pH 8.5; 25 °C; 1.55 bar;	Lignin (95%); vanillin ($\approx$0)
	TiO_2	5	1.2 m/s; C_{feed} 60/5 g L^{-1} lignin/vanillin; pH 12.5; 25 °C; 1.55 bar	Lignin (97%); vanillin ($\approx$0)
	Al_2O_3-TiO_2	15	0.992 m/s; C_{feed} 60/6 g L^{-1} lignin/vanillin; pH 12.5; 25 °C; 1.55 bar	Lignin (87%); vanillin ($\approx$0)
			0.992 m/s; C_{feed} 60/6 g L^{-1} lignin/vanillin; pH 8.5; 25 °C; 1.55 bar	Lignin (94%); vanillin ($\approx$0)
Oxidized *Eucalyptus* industrial kraft liquor[a]	ZrO_2	50	1.2 m/s and C_{feed} with 86.5 g/L of TS and 2.40 g/L of TP; pH 10.1; 25 °C; 1.4 bar	TS (25%) TP ($\approx$0)
	TiO_2	5	1.2 m/s and C_{feed} with 75.3 g/L of TS and 2.38 g/L of TP; pH 10.1; 25 °C; 1.4 bar	TS (29%) TP (11.5%)
	TiO_2	1	1.2 m/s and C_{feed} with 64.0 g/L of TS and 2.22 g/L of TP; pH 10.1; 25 °C; 1.4 bar	TS (15%) TP (9.0%)

PDMS polydimethylsiloxane, *PAN* polyacrylonitrile, *TS* total nonvolatile solids, *TP* total phenolate compounds quantified by HPLC-UV (*p*-hydroxybenzaldehyde, vanillin, syringaldehyde, vanillic acid, syringic acid, acetovanillone, and acetosyringone
[a]solution diluted 3x and enriched in some phenolic compounds, it corresponds to a fractionation sequence 50kDa→ 5 kDa→1kDa

regarding the molecular weight by gel permeation chromatography (GPC), total non-volatile solids, and total phenolate compounds quantified by HPLC-UV.

The 50 kDa membrane did not retain vanillate and any other lower molecular weight phenolates quantified. On the other hand, contrary to what was observed in Žabková et al. 2007 study (Žabková et al. 2007b), these low molecular weight compounds were retained in the 5 and 1 kDa membrane stages, with retention coefficients of 11.5% and 9%, respectively. The observed rejection coefficients are explained by the greater complexity of the feed mixture employed by Mota et al. (2017) probably explained by different interactions of solutes with other molecules or with the membrane, generating different fouling mechanisms.

Moreover, total solid rejection coefficients observed were between 15% and 29%. This trend is explained by the fact that the total solids quantified correspond to a measure of all compounds present in the depolymerized mixture ranging from the highest to the lowest molecular weight compounds present in the feed mixture. GPC helped in understanding the influence of each membrane stage in the MW profiles of retentate and permeate streams. The retentate streams got richer in the higher molecular weight compounds, while the permeate streams got more depleted.

Mota et al. (2017) study encompassed the analysis of the contribution of fouling on the membrane productivity observed by a resistance-in-series approach showing that the 1 kDa membrane was the one more affected by fouling. The reversible fouling was the component more relevant in 1 and 50 kDa processing, while for the 5 kDa, the irreversible fouling component had a somewhat higher contribution.

Cleaning efficiency was evaluated for each membrane stage employing 0.1 M NaOH solution and similarly to what was obtained by Žabková et al. In 2007 (Žabková et al. 2007b) study, the initial flux was recovered, indicating that the ceramic membranes employed to treat the real lignin oxidized solution can be a viable solution to implement in an industrial scale.

2.2.2 Ion-Exchange Process for Vanillin Recovery

Ion-exchange studies in the perspective of recovering phenolic compounds from alkaline oxidized lignin solutions have been developed employing mostly cationic ion-exchange resins (Fargues et al. 1996a; Forss et al. 1986; Logan 1965; Toppel 1959; Žabková et al. 2007a). The main focus is on vanillin recovery from synthetic mixtures prepared in NaOH or alkaline oxidized lignin solutions.

The cation-exchange processes are applied to neutralize the compounds and, therefore, avoid the consumption of large amounts of acid that would be required in a direct acidification of the oxidized mixture. According to Forss et al. (1986), about 60% of acid can be saved by replacing the acid neutralization step by ion-exchange processes. Nevertheless, previous separation (e.g., by ultrafiltration) of the polymeric/oligomeric lignin fragments is mandatory in order to avoid precipitation problems in the ion-exchange system.

Toppel (1959) has patented the first ion-exchange process focused on vanillin adsorption enhancing the interaction between the carbonyl group of the vanillin molecule through the ammonia group previously introduced by charging the polystyrene sulfonic acid cation exchanger with hydroxylammonium chloride solution. The vanillin was recovered employing a 3 M hydrochloric acid solution.

Forss et al. (1986) reported the separation of sodium vanillate from lignosulfonates, sodium hydroxide, and sodium carbonate employing several cation-exchange resins. Water or sodium carbonate solutions were employed as eluting agents. The separation is accomplished due to the different elution orders and yields vanillin recoveries above 73%. Sodium carbonate solutions have influenced slightly the oxidized lignin elution order with the advantage of increasing the adsorption of the oxidized products. However, the elution volume needed is higher. In this process, lignin and sodium can be reintroduced in the processes of the pulp mill, while the enriched solution containing vanillin is neutralized with sulfuric acid employing less 78% of the acid solution needed to neutralize the oxidized spent sulfite liquor.

Logan (1965) studied two weak cation-exchange resins Amberlite IRC-50 and Duolite C63 with carboxylic and phosphoric acid exchange centers, respectively. The main focus was the sodium recovery from an alkaline oxidized lignin medium in order to be reused later. In this process, sodium vanillate and other phenolates present in the alkaline oxidized lignin medium are converted into their protonated form.

Fargues et al. (1996a) and Žabková et al. (2007a) have demonstrated the possibility of employing the strong acid cationic exchange resins Duolite C20 and Amberlite IRA120H, respectively, to recover vanillin in its sodium form from synthetic mixtures. The counter ions employed were different, Na^+ and H^+ for the first and second resins, respectively.

Žabková et al. (2007a) have develop a mathematical model to explain the ion-exchange process proposed with simultaneous neutralization. The effect of the feed solution pH was also evaluated ranging from 8.2 to 12.1 where it was observed that the higher the pH of the feed solution, the lower was the breakthrough time of vanillate. The expected rectangular behavior of the isotherm in the ion-exchange process is affected in the presence of a vanillin/vanillate buffer system.

Anion-exchange resins have been applied by Stecker et al. (2014) and Schmitt et al. (2015) to treat an electrochemically oxidized lignin medium in order to recover vanillin and other oxidation products produced. The ion-exchange processes suggested have the advantage of being used in process intensification since it can be performed simultaneously with the electrochemical oxidation reaction. The oxidation products can be eluted employing 2% hydrochloric acid in methanol or ethyl acetate: acetic acid (80:20, v/v). It is important to note that when employing complex mixtures such as the case of real oxidized lignin solutions, further purification methods are needed in order to obtain purified fractions of vanillin or any other phenolic compound present (e.g., Sy, acetovanillone, guaiacol, vanillic acid).

2.2.3 Adsorption and Desorption of Vanillin and Syringaldehyde onto Polymeric Resins

Adsorption studies aiming the recovery of typical phenolic monomers found in lignin oxidized media have been performed with several polymeric resins employing either synthetic model solutions (Huang et al. 2013; Jin and Huang 2013; Mota et al. 2016a; Samah et al. 2013; Xiao et al. 2009; Žabková et al. 2006; Zhang et al. 2008) or real complex oxidized lignin solutions (Gomes et al. 2018; Mota 2017; Wang et al. 2010).These studies have demonstrated the suitability of this type of resins to recover the desired monomers with great adsorptive capacities. Satisfactory mathematical models to describe adsorption of vanillin, syringaldehyde, and respective acids have been developed (Mota et al. 2016a; Žabková et al. 2006). Moreover, the application of this type of resins to continuously recover produced vanillin from fermentation broths has also been successfully demonstrated (Hua et al. 2007; Nilvebrant et al. 2001; Stentelaire et al. 2000; Wang et al. 2005; Zhao et al. 2006). Table 2.7 summarized the main studies encompassing synthetic mixtures prepared in aqueous media mostly referring to vanillin. It is important to note that literature studies refer to specific experimental conditions and comparison of the maximum adsorption capacities listed should be done carefully. Among the literature data, the highest adsorption capacities obtained for vanillin refer to the resins SP700 and XAD16N of 663 and 587 mg $g^{-1}_{dry\ resin}$, respectively. Vanillic and syringic acid adsorption studies onto SP700 resin have shown that the phenolic acids are adsorbed in a lesser extent than their respective aldehydes.

Many of the studies employing synthetic mixtures encompass the desorption of the phenolic compounds by different eluents. Mota et al. (2016a) have shown that about 83–85% of vanillin or syringaldehyde can be desorbed with water; however, considerable bed volumes of water are employed resulting in very diluted final solutions, and therefore, developing an industrial process employing water must be carefully evaluated and will depend on the next purification steps.

On the other hand, organic solvents could be more attractive as eluting agents since they can readily recover the phenolic compounds adsorbed while simultaneously concentrating the solution (Mota 2017; Mota et al. 2016b; Wang et al. 2005, 2010; Xiao et al. 2009). This behavior is explained by the Hildebrand solubility parameter (Hildebrand and Scott 1950) of "like likes like"; the phenolic solutes having higher solubility in organic solvents than water will be preferably adsorbed when dissolved in water and desorbed with organic solvent eluents.

Desorption studies have shown that ethanol and ethanol aqueous solutions above 40%v/v can recover more than 70% of the adsorbed compounds and that adding small amounts of acid does not enhance the desorption (Mota et al. 2016b; Xiao et al. 2009). Additionally, it was demonstrated that about 83% of vanillin, syringaldehyde, or respective acids can be readily eluted with very few bed volumes of ethanol/water 90:10%v/v solution achieving simultaneously a high level of concentration (Mota 2017; Mota et al. 2016b). In accordance, Zhou and Wang (Zhou and Wang 2008) accomplished full vanillin recovery with 4.6 bed volumes of absolute ethanol.

Table 2.7 Maximum adsorption capacities of typical phenolic compounds (PC) present in oxidized lignin reaction media employing polymeric resins and monocomponent synthetic mixtures

	Resin	Conditions			Adsorption capacity (mg g^{-1})	Ref
		T (K)	pH	Ce_{max} (mg L^{-1})		
V	SP206[a]	293–333	5.3	53	115	Žabková et al. (2006)
	SP206[b]	293	6.5	91	≈ 91–106	Žabková et al. (2006)
	H103[a]	293	6.0	400–700	416	Zhang et al. (2008)
	H103[a]	293–328	–	30	73	Samah et al. (2013)
	HJ-J08[a]	300–320	–	450	338,358	Jin and Huang (2013)
	Polydivinylbenzene/ polyacrylethylenediamine interpenetrating polymer networks[a]	293–308	–	–	73–103	Xiao et al. (2009)
	SP700[a]	283–313		300–500	663	Mota et al. (2016a)
	XAD16N[a]	283–313		450–500	587	Mota et al. (2016a)
Sy	SP700[a]	288–313		150–400	707	Mota et al. (2016a)
	XAD16N[a]	288–313		200–300	730	Mota et al. (2016a)
VA	SP700[a]	288–313		150–200	450	Mota (2017)
SA	SP700[a]	288–313		70	423	Mota (2017)
Hy	Resorcinol-modified poly(styrene-co-divinylbenzene)[a]	298	–	500	246–282	Huang et al. (2013)

Vanillin (V), syringaldehyde (Sy), vanillic acid (VA), syringic acid (SA), p-hydroxybenzaldehyde (Hy)
[a]water medium
[b]NaOH/water medium

This constitutes a great advantage considering a next processing step of purification (e.g., crystallization or spray drying) since less volume needs to be treated and ethanol is easily removed.

When employing real solutions, it is expected changes in the adsorptive behavior of the polymeric resins since several compounds present in solution will compete for the same adsorption sites of the resin. In order to understand the behavior of the polymeric resins to adsorb vanillin and the other phenolic compounds of interest in the presence of real and complex oxidized lignin media, it has been increasing the number of studies with real oxidized lignin feed solutions (Gomes et al. 2018; Mota 2017; Wang et al. 2010; Wu et al. 2003). Some of the studies are listed in Table 2.8.

Table 2.8 Adsorption capacities of phenolic compounds in oxidized lignin reaction media employing polymeric resins and real oxidized mixtures (Mota 2017; Wang et al. 2010)

| Medium | Resin | Target compounds | Conditions | | | Adsorption capacity (mg g^{-1}) |
			T (K)	pH	Ce (mg L^{-1})	
Oxidized IKL previously submitted to UF	SP700	V	298	≈ 7	44	79.7
		Sy			39	67.3
		VA			43	9.2
		SA			48	5.4
		VO			13	30.9
		SO			24	55.7
Oxygen delignification liquor	D101	V and Sy	290	4.5	6.1–7.3	4.1–4.9

Vanillin (V), syringaldehyde (Sy), vanillic acid (VA), syringic acid (SA), acetovanillone (VO), and acetosyringone (SO)

Wang et al. (2010) were successful in demonstrating the recovery of vanillin and syringaldehyde from an oxygen delignification liquor employing the polymeric resin D101. Desorption studies with ethyl ether managed to recover almost all adsorbed vanillin and syringaldehyde (94–96%) with 1.3 bed volumes achieving a final solution containing 37.5% vanillin, 31.9% syringaldehyde, and 30.6% of other compounds (mainly acetosyringone).

A recent study deals with loading a real oxidized industrial Eucalyptus kraft liquor solution onto the SP700 resin (Mota 2017). The oxidized solution was previously submitted to a membrane fractionation sequence with 50, 5, and 1 kDa membranes and correction of the pH value to 7.5–8 with H_2SO_4 in order to convert the ionized phenolic compounds into their neutral state, thus favoring their adsorption onto the polymeric resin.

The complete saturation of the bed was performed with the final permeate stream obtained from the membrane fractionation sequence and the maximum adsorption capacities for several monomers assessed, being summarized in Table 2.8. As expected, the adsorption capacities were smaller than the ones determined from the monocomponent studies performed with synthetic solutions due to competition of several phenolic monomers and other compounds present for the adsorption sites of the resin. In this study, the authors have observed that the vanillic and syringic acids are practically not adsorbed at the studied pH value, explained by their lower pKa value around 4.3–4.4 (Ragnar et al. 2000). About 83% of vanillin, syringaldehyde, and respective ketones were recovered with ethanol:water 90:10%v/v solution. Starting with a feed solution containing 0.83 g/L of vanillin and syringaldehyde, a final enriched solution with 5.31 g/L was obtained after desorption with aqueous ethanol.

In the same study, the capacity of the resin to be regenerated and reused was also assessed employing the permeates obtained from 5 to 1 kDa membrane stages. It was demonstrated that SP700 resin can be reutilized up to 4–5 cycles of adsorption/desorption without losing performance in adsorption of the phenolic compounds of interest, constituting a key factor for developing a sustainable industrial adsorption step. Each cycle encompassed a first stage with 10 min of adsorption (not achieving saturation of the bed and, thus, avoiding losing vanillin and syringaldehyde at column outlet), a washing step with water for 1 min, a desorption step with aqueous ethanol for 4 min, and regeneration step with water for 30 min (Mota 2017).

Several studies have demonstrated that the pH value is an important parameter to take into consideration when employing polymeric resins and of the importance of knowing the acid dissociation constant of the target compounds (Gomes et al. 2018; Wang et al. 2010; Žabková et al. 2006). Generally, it is expected that the adsorptive capacity of the polymeric resins for phenolic compounds decreases with the pH value increase due to the fact that the ionized compound fraction becomes more relevant and is not adsorbed.

Aiming to understand the effect of the sodium hydroxide in vanillin adsorption from alkaline oxidized solution, Žabková et al. (2006) prepared synthetic mixtures of vanillin in different concentrations of sodium hydroxide solutions and concluded that the adsorption capacity is negatively affected by the sodium hydroxide or, in other words, by the increase of pH value from 6.5 to 9.5 due to the fact that with the increasing pH value, most of the vanillin is in vanillate form and is not adsorbed.

Wang et al. (2010) also demonstrated that increasing the pH value of an oxygen delignification liquor from 4.5 to 6.5, vanillin and syringaldehyde adsorption capacity onto D101 resin decreases.

Typical phenolic compounds present in oxidized media have different acid dissociation constants (pKa), and therefore the separation among aldehydes, acids, and ketones can be somewhat tuned by the pH of the feed solution due to acid dissociation constant differences. Mota et al. (2017) have shown that loading onto SP700 resin, an oxidized industrial kraft liquor containing several families of compounds at pH 7, the vanillic and syringic acids are practically not adsorbed onto the resin, and separation from these acids and the respective ketones (acetovanillone and acetosyringone) and aldehydes (vanillin and syringaldehyde) can be accomplished.

Gomes et al. (2018) have shown that tuning the pH of the oxidized lignin feed solution and sequential elution with water and absolute ethanol can make possible the separation between families of phenolic compounds present in solution: aldehydes (vanillin and syringaldehyde), acids (vanillic and syringic acids), and ketones (acetovanillone and acetosyringone). This process can simplify the current laborious separation and purification of oxidized lignin medium solutions employing several liquid-liquid extractions with harmful solvents and precipitation (Mota et al. 2016c). The choice of the operational pH value will depend on the final applications and purification processes to be applied.

The authors loaded an oxidized organosolv lignin from tobacco stalks previously ultrafiltrated with a 5 kDa MW cutoff membrane at different pH values ranging from 9 to 12. At these pH ranges, the phenolic acids are not adsorbed and can be collected at the outlet of the column during feeding step. The aldehydes and ketones are collected during desorption steps with water and absolute ethanol, respectively. According to the feeding pH value, different amounts of vanillin, syringaldehyde, acetovanillone, and acetosyringone are desorbed in the aqueous and ethanol phases. Generally, increasing the pH value of the feed solution, the amount of aldehydes and ketones recovered in the aqueous phase increases but at different desorption rates: for pH values above 10, more than 80% of the aldehydes and 30–50% of the ketones are recovered.

On the contrary, the amounts of aldehydes and ketones recovered in the ethanol eluting step decrease with pH increase. However, similar to that observed during elution with water, different amounts of aldehydes and ketones are recovered: the amount of the aldehydes recovered for pH above 10 is less than 10%, and about 30% and 60% of the ketones are recovered.

Concluding, the feeding pH value should be selected according to the goal to be achieved: maximum purity degree, maximum amount of mass recovered, or further purification step to be applied.

2.3   The Integrated Process for Complete Lignin Valorization into Phenolic Compounds and Polyurethanes

The integrated oxidation and separation process for lignin valorization in pulp and paper streams was proposed by Borges da Silva et al. (2009) and joins two distinct lignin valorization routes: a first one encompassing the chemical breakdown of the three- dimensional network structure of lignin to obtain value-added phenolic monomers and a second one, completely opposite to the first one, which takes advantage of the lignin functionalities and considers the production of more complex matrices and bio-based polymers. In Fig. 2.7, it presented the flow sheet of the integrated process suggested. This proposal is based on the extensive research and know-how of the Laboratory of Separation and Reaction Engineering (LSRE) team concerning the lignin oxidation studies (batch and continuous systems) (Araújo et al. 2009; Fargues et al. 1996a; b; Mathias et al. 1995b; Mathias and Rodrigues 1995), the separation studies by ultrafiltration (Žabková et al. 2007b) and chromatography processes (Zabka et al. 2006; Žabková et al. 2007a, b; Žabková et al. 2006), and the studies regarding lignin-based polymers synthesis (Cateto et al. 2008). The different approaches have been detailed above including the recent research developments of the LSRE group that also support the integrated process suggested.

In the integrated process, the black liquor is initially processed in order to isolate lignin by known methods (e.g., Lignoboost, acid precipitation, or ultrafiltration). The purified lignin is oxidized in alkaline conditions with O_2 (Araújo 2008). The complex mixture containing the phenolate monomers and fragmented lignin is subject to UF where the high MW lignin fragments are retained (retentate stream) while the phenolate monomers and other low MW compounds go to the permeate stream (Žabková et al. 2007b).

The permeate stream is loaded into a column packed with an H^+ form resin, and simultaneous neutralization and reaction occur, where vanillate and other phenolate species are converted to vanillin and phenolics in their protonated form (Žabková

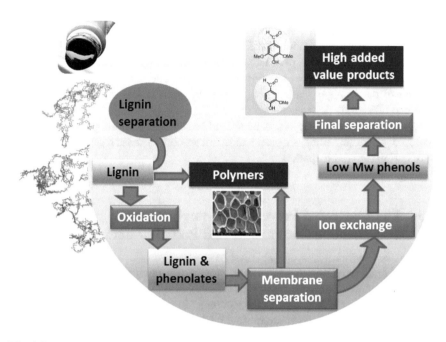

Fig. 2.7 Flow sheet of the integrated process in an industrial unit for valorization of process streams from lignocellulosic-based biorefineries

et al. 2007a). The ultrafiltration step is important to avoid precipitation problems in this step, and therefore, the MW cutoff of the membrane separation system should be carefully selected.

The ion-exchange process is a good alternative to the currently available methods, the direct neutralization employing acids (Hilbbert and Tomlinson 1937) for posterior liquid-liquid extraction with benzene or toluene or the direct liquid-liquid extraction with butyl or isopropyl alcohol (Bryan 1955; Sandborn et al. 1936). The first methodology has the disadvantage of consuming large amounts of acid to precipitate lignin, and in the process, the oxidation products are also dragged with lignin. The second process is limited by the lower extraction yield due to low solubility of the oxidation products in the alcohol. Moreover, both conventional processes employ harmful organic solvents.

The retentate stream can be either employed to supply energy to the process or produce lignin-based PU products (Cateto et al. 2008) such as rigid foams, elastomers, or sealants.

This integrated process designed emerges as the required key piece for the sustainable production of value-added compounds and material from lignin. It has the clear advantage of avoiding the often laborious separation and purification processes encompassing several extraction and precipitation steps with harmful solvents to obtain the phenolic monomers that have already been summarized by Mota et al. (2016c). Moreover, two distinct valorization routes of lignin are joined in an integrated process being very attractable in a point of view of developing sustainable lignocellulosic biorefineries since more value can be made out of the same biomass.

References

Almécija MC, Ibáñez R, Guadix A, Guadix EM (2007) Effect of pH on the fractionation of whey proteins with a ceramic ultrafiltration membrane. J Membr Sci 288:28–35

Araújo JD (2008) Production of vanillin from lignin present in the Kraft black liquor of the pulp and paper industry. PhD thesis, University of Porto

Araújo JDP, Grande CA, Rodrigues AE (2009) Structured packed bubble column reactor for continuous production of vanillin from kraft lignin oxidation. Catal Today 147., Supplement:S330–S335

Araújo JDP, Grande CA, Rodrigues AE (2010) Vanillin production from lignin oxidation in a batch reactor. Chem Eng Res Des 88:1024–1032

Arkell A, Olsson J, Wallberg O (2014) Process performance in lignin separation from softwood black liquor by membrane filtration. Chem Eng Res Des 92:1792–1800

Azadi P, Inderwildi OR, Farnood R, King DA (2013) Liquid fuels, hydrogen and chemicals from lignin: a critical review. Renew Sust Energ Rev 21:506–523

Behling R, Valange S, Chatel G (2016) Heterogeneous catalytic oxidation for lignin valorization into valuable chemicals: what results? What limitations? What trends? Green Chem 18:1839

Benfer S, Árki P, Tomandl G (2004) Ceramic membranes for filtration applications – preparation and characterization. Adv Eng Mater 6:495–500

Borges da Silva EA, Žabková M, Araújo JD, Cateto CA, Barreiro MF, Belgacem MN, Rodrigues AE (2009) An integrated process to produce vanillin and lignin-based polyurethanes from kraft lignin. Chem Eng Res Des 87:1276–1292

Bryan CC (1955) Propanol extraction of sodium vanillinate. US Patent 2721221

Cateto CA, Barreiro MF, Rodrigues AE (2008) Monitoring of lignin-based polyurethane synthesis by FTIR-ATR. Ind Crop Prod 27:168–174

Christopher L (2012) Integrated forest biorefineries: challenges and opportunities. Royal Society of Chemistry, Cambridge

Collis BC (1954) Manufacture of vanillin from lignin. United States Patent US2692291A,

Costa CAE, Pinto PCR, Rodrigues AE (2015) Radar tool for lignin classification on the perspective of its valorization. Ind Eng Chem Res 54:7580–7590

de la Rubia Á, Rodríguez M, Prats D (2006) pH, Ionic strength and flow velocity effects on the NOM filtration with TiO_2/ZrO_2 membranes. Sep Purif Technol 52:325–331

Dardelet S, Froment P, Lacoste N, Robert A (1985) Aldéhyde syringique: Possibilités de production à partir de bois feuillus. Revue – ATIP 39:267–274

Dong DJ, Fricke AL, Moudgil BM, Johnson H (1996) Electrokinetic study of kraft lignin. TAPPI J 79:191

Erofeev YV, Afanas'eva VL, Glushkov RG (1990) Synthetic routes to 3,4,5-trimethoxybenzaldehyde (review). Pharm Chem J 24:501–510

Evju H (1979) Process for preparation of 3-methoxy-4-hydroxybenzaldehyde. US Patent, p 4151207

Fargues C, Mathias ÁL, Rodrigues A (1996a) Kinetics of vanillin production from kraft lignin oxidation. Ind Eng Chem Res 35:28–36

Fargues C, Mathias ÁL, Silva J, Rodrigues A (1996b) Kinetics of vanillin oxidation. Chem Eng Technol 19:127–136

Forss KG, Talka ET, Fremer KE (1986) Isolation of vanillin from alkaline oxidized spent sulfite liquor. Ind Eng Chem Prod Res Dev 25:103–108

German Federal Government (2012) Biorefineries roadmap. Federal Ministry of Food, Agriculture and Consumer Protection, Berlin, https://www.bmbf.de/pub/Roadmap_Biorefineries_eng.pdf. Acessed 17 Sep 2018

Gierer J (1986) Chemistry of delignification. Wood Sci Technol 20:1–33

Gomes ED, Mota MI, Rodrigues AE (2018) Fractionation of acids, ketones and aldehydes from alkaline lignin oxidation solution with SP700 resin. Sep Purif Technol 194:256–264

Hilbbert H, Tomlinson GHJ (1937) Manufacture of vanillin from waste sulphite pulp liquor United States Patent US Patent 2069185

Hildebrand JH, Scott RL (1950) Solubility of Non-Electrolytes. Reinhold, New York

Hocking MB (1997) Vanillin: synthetic flavoring from spent sulfite liquor. J Chem Educ 74:1055

Holladay JE, Bozell JJ, White JF, Johnson D (2007) Top value-added chemicals from biomass. Volume II – Results of screening for potential. Candidates from Biorefinery Lignin. PNNL-16983

Hua D, Ma C, Song L, Lin S, Zhang Z, Deng Z, Xu P (2007) Enhanced vanillin production from ferulic acid using adsorbent resin. Appl Microbiol Biotechnol 74:783–790

Huang J, Yang L, Zhang Y, Pan C, Liu Y-N (2013) Resorcinol modified hypercrosslinked poly(styrene-co-divinlybenzene) resin and its adsorption equilibriums, kinetics and dynamics towards p-hydroxylbenzaldehyde from aqueous solution. Chem Eng J 219:238–244

Ibrahim MNM, Sriprasanthi RB, Shamsudeen S, Adam F, Bhawani SA (2012) A concise review of the natural existance, synthesis, properties, and applications of syringaldehyde. Bioresources 7(3):4377–4399

Jin X, Huang J (2013) Adsorption of vanillin by an anisole-modified hyper-cross-linked polystyrene resin from aqueous solution: equilibrium, kinetics, and dynamics. Adv Polym Technol 32:E221–E230

Jönsson A-S, Nordin A-K, Wallberg O (2008) Concentration and purification of lignin in hardwood kraft pulping liquor by ultrafiltration and nanofiltration. Chem Eng Res Des 86:1271–1280

Lin SY, Dence CW (1992) Methods in lignin chemistry. Springer-Verlag, Berlin

Logan CD (1965) Cyclic process for recovering vanillin and sodium values from lignosulfonic waste liquors by ion exchange. US Patent 3197359

Ma R, Xu Y, Zhang X (2015) Catalytic oxidation of biorefinery lignin to value-added chemicals to support sustainable biofuel production. Chem Sus Chem 8:24–51

Mahmood N, Yuan Z, Schmidt J, Xu C (2016) Depolymerization of lignins and their applications for the preparation of polyols and rigid polyurethane foams: a review. Renew Sust Energ Rev 60:317–329

Manchand PS, Rosen P, Belica PS, Oliva GV, Perrotta AV, Wong HS (1992) Syntheses of antibacterial 2,4-diamino-5-benzylpyrimidines. Ormetoprim and trimethoprim. J Org Chem 57:3531–3535

Marshall HB, Sankey AC (1951) Method of producing vanillin. United States Patent 2544999

Mathias AL (1993) Produção de vanilina a partir da lenhina: Estudo cinético e do processo (in Portuguese language). PhD thesis, Faculty of Engineering University of Porto

Mathias AL, Rodrigues AE (1995) Production of vanillin by oxidation of Pine kraft lignins with oxygen. Holzforsch. – Int J Biol Chem Phys Technol Wood 49:273–278

Mathias AL, Lopretti MI, Rodrigues AE (1995a) Chemical and biological oxidation of Pinus-Pinaster lignin for the production of vanillin. J Chem Technol Biot 64:225–234

Mathias ÁL, Lopretti MI, Rodrigues AE (1995b) Chemical and biological oxidation of Pinus pinaster lignin of the production of vanillin. J Chem Technol Biotechnol 64:225–234

Mota IF (2017) Fractionation and purification of syringaldehyde and vanillin from oxidation of lignin. PhD thesis, Faculty of Engineering of University of Porto

Mota MIF, Pinto PCR, Loureiro JM, Rodrigues AE (2016a) Adsorption of vanillin and syringaldehyde onto a macroporous polymeric resin. Chem Eng J 288:869–879

Mota MIF, Pinto PCR, Loureiro JM, Rodrigues AE (2016b) Successful recovery and concentration of vanillin and syringaldehyde onto a polymeric adsorbent with ethanol/water solution. Chem Eng J 294:73–82

Mota MIF, Rodrigues Pinto PC, Loureiro JM, Rodrigues AE (2016c) Recovery of vanillin and syringaldehyde from lignin oxidation: a review of separation and purification processes. Sep Purif Rev. 45:227–259

Nilvebrant N-O, Reimann A, Larsson S, Jönsson L (2001) Detoxification of lignocellulose hydrolysates with ion-exchange resins. Appl Biochem Biotechnol 91–93:35–49

Pandey MP, Kim CS (2011) Lignin depolymerization and conversion: a review of thermochemical methods. Chem Eng Technol 34:29–41

Pinto PC, Borges da Silva EA, Rodrigues AE (2011) Insights into oxidative conversion of lignin to high-added-value phenolic aldehydes. Ind Eng Chem Res 50:741–748

Pinto PCR, Borges da Silva EA, Rodrigues AE (2012) Lignin as source of fine chemicals: vanillin and syringaldehyde. In: Baskar C, Baskar S, Dhillon RS (eds) Biomass conversion. Springer, Berlin, Heidelberg, pp 381–420

Pinto PCR, Costa CE, Rodrigues AE (2013) Oxidation of lignin from *Eucalyptus globulus* pulping liquors to produce syringaldehyde and vanillin. Ind Eng Chem Res 52:4421–4428

Ragnar M, Lindgren CT, Nilvebrant N-O (2000) *pKa*-values of guaiacyl and syringyl phenols related to lignin. J Wood Chem Technol 20:277–305

Sales FG, Maranhão LCA, Lima Filho NM, Abreu CAM (2006) Kinetic evaluation and modeling of lignin catalytic wet oxidation to selective production of aromatic aldehydes. Ind Eng Chem Res 45:6627–6631

Samah RA, Zainol N, Yee PL, Pawing CM, Abd-Aziz S (2013) Adsorption of vanillin using macroporous resin H103. Adsorpt Sci Technol 31:599–610

Sandborn LT, Salvesen JR, Howard GC (1936) Process of making vanillin. US Patent 2057117

Santos SG, Marques AP, Lima DLD, Evtuguin DV, Esteves VI (2011) Kinetics of eucalypt lignosulfonate oxidation to aromatic aldehydes by oxygen in alkaline medium. Ind Eng Chem Res 50:291–298

Schmitt D, Regenbrecht C, Hartmer M, Stecker F, Waldvogel SR (2015) Highly selective generation of vanillin by anodic degradation of lignin: a combined approach of electrochemistry and product isolation by adsorption. Beilstein J Org Chem 11:473–480

Schutyser W, Renders T, Van den Bosch S, Koelewijn SF, Beckham GT, Sels BF (2018) Chemicals from lignin: an interplay of lignocellulose fractionation, depolymerisation, and upgrading. Chem Soc Rev 47:852–908

Sippola VO, Krause AOI (2005) Bis(o-phenanthroline)copper-catalysed oxidation of lignin model compounds for oxygen bleaching of pulp. Catal Today 100:237–242

Sjöström E (1993) Wood chemistry, fundamentals and applications. Academic Press, Inc., San Diego, California

Sridhar P, Araujo JD, Rodrigues AE (2005) Modeling of vanillin production in a structured bubble column reactor. Catal Today 105:574–581

Stecker F et al (2014) Method for obtaining vanillin from aqueous basic compositions containing vanillin. WO Patent 2014006108 A1

Stentelaire C, Lesage-Meessen L, Oddou J, Bernard O, Bastin G, Ceccaldi BC, Asther M (2000) Design of a fungal bioprocess for vanillin production from vanillic acid at scalable level by *Pycnoporus cinnabarinus*. J Biosci Bioeng 89:223–230

Sundin J (2000) Precipitation of kraft lignin under alkaline conditions. Ph.D thesis, Royal Institute of Technology, Stockholm

Tarabanko V, Petukhov D (2003) Study of mechanism and improvement of the process of oxidative cleavage of lignins into the aromatic aldehydes. Chem Sustain Dev 11:655–667

Tarabanko VE, Pervishina EP, Hendogina YV (2001) Kinetics of aspen wood oxidation by oxygen in alkaline media. React Kinet Catal Lett 72:153–162

Tarabanko VE, Petukhov DV, Selyutin GE (2004) New mechanism for the catalytic oxidation of lignin to vanillin. Kinet Catal 45:569–577

Toledano A, García A, Mondragon I, Labidi J (2010a) Lignin separation and fractionation by ultrafiltration. Sep Purif Technol 71:38–43

Toledano A, Serrano L, Garcia A, Mondragon I, Labidi J (2010b) Comparative study of lignin fractionation by ultrafiltration and selective precipitation. Chem Eng J 157:93–99

Toppel O (1959) Method for the separation of carbonyl compounds. US Patent 2897238

Tromans D (1998) Oxygen solubility modeling in inorganic solutions: concentration, temperature and pressure effects. Hydrometallurgy 50:279–296

Uloth VC, Wearing JT (1988) Kraft lignin recovery: acid precipitation versus ultrafiltration. I laboratory test results. Pulp Paper Can 90:67–71

Villar JC, Caperos A, Garcia-Ochoa F (2001) Oxidation of hardwood kraft-lignin to phenolic derivatives with oxygen as oxidant. Wood Sci Technol 35:245–255

Wang C-l, S-l LI, Zhou Q-l, Zhang M, Zeng J-h, Zhang Y (2005) Extraction of vanillin in fermented broth by macroporous adsorption resin. Jingxi Huagong (Fine Chemicals) 22:458–460

Wang Z, Chen K, Li J, Wang Q, Guo J (2010) Separation of vanillin and syringaldehyde from oxygen delignification spent liquor by macroporous resin adsorption. Clean (Weinh) 38:1074–1079

Werhan H, Farshori A, Rudolf von Rohr P (2012) Separation of lignin oxidation products by organic solvent nanofiltration. J Membr Sci 423–424:404–412

Wong Z, Chen K, Li J (2010a) Formation of vanillin and syringaldehyde in an oxygen delignification process. Bioresour Technol 5:1509–1516

Wong Z, Chen K, Li J (2010b) Formation of vanillin and syringaldehyde in an oxygen delignifiction process. Bioresources 5(3):1509–1516

Wu G, Heitz M (1995) Catalytic mechanism of Cu^{2+} and Fe^{3+} in alkaline O_2 oxidation of lignin. J Wood Chem Technol 15:189–202

Wu G, Heitz M, Chornet E (1994) Improved alkaline oxidation process for the production of aldehydes (vanillin and syringaldehyde) from steam-explosion hardwood lignin. Ind Eng Chem Res 33:718–723

Wu X, Zhang N-z, Qi C-h (2003) Enrichment of vanillin with adsorbent resin in the oxidative liquorsof acidic sulphite pulping, vol 24. China Pulp & Paper Industry, Tianjin, China, pp. 21

Xiao G-Q, Xie X-L, Xu M-C (2009) Adsorption performances for vanillin from aqueous solution by the hydrophobic – hydrophilic macroporous polydivinylbenzene/polyacrylethylenediamine IPN resin. Acta Phys -Chim Sin 25:97–102

Zabka M, Minceva M, Rodrigues AE (2006) Experimental and modeling study of adsorption in preparative monolithic silica column. Chem Eng Process Process Intensif 45:150–160

Zakzeski J, Bruijnincx PCA, Jongerius AL, Weckhuysen BM (2010) The catalytic valorization of lignin for the production of renewable chemicals. Chem Rev 110:3552–3599

Zhang Q-F, Jiang Z-T, Gao H-J, Li R (2008) Recovery of vanillin from aqueous solutions using macroporous adsorption resins. Eur Food Res Technol 226:377–383

Zhang J, Deng H, Lin L (2009) Wet aerobic oxidation of lignin into aromatic aldehydes catalysed by a perovskite-type oxide: $LaFe_{1-x}Cu_xO_3$ (x=0, 0.1, 0.2). Molecules 14:2747–2757

Zhao L-Q, Sun Z-H, Zheng P, He J-Y (2006) Biotransformation of isoeugenol to vanillin by *Bacillus fusiformis* CGMCC1347 with the addition of resin HD-8. Process Biochem 41:1673–1676

Zhou Q-l, Wang H-l (2008) Study on separation and purification of total vanillin by adsorbing resin. Acta Agriculturae Jiangxi 28:34

Žabková M, Otero M, Minceva M, Zabka M, Rodrigues AE (2006) Separation of synthetic vanillin at different pH onto polymeric adsorbent Sepabeads SP206. Chem Eng Process Process Intensif 45:598–607

Žabková M, Borges da Silva EA, Rodrigues AE (2007a) Recovery of vanillin from Kraft lignin oxidation by ion-exchange with neutralization. Sep Purif Technol 55:56–68

Žabková M, Borges da Silva EA, Rodrigues AE (2007b) Recovery of vanillin from lignin/vanillin mixture by using tubular ceramic ultrafiltration membranes. J Membr Sci 301:221–237

Chapter 3
Polyurethanes from Recovered and Depolymerized Lignins

Abstract This chapter discusses lignin potential as a raw material for polyurethane synthesis having in view the valorization of a coproduct generated by the pulp and paper industry, bioethanol production, and other biorefinery processes. It addresses the use of lignin as such, i.e., without chemical modification, and the use of lignin after chemical modification. The later encompasses the description of some liquefaction processes and oxypropylation. The advantages of using depolymerized lignins in the context of polyurethanes will be also addressed. Different lignin-based polyurethane systems, e.g., rigid polyurethane foams, flexible polyurethane foams, lignin-based composites, and thermoplastic polyurethanes, obtained from different lignin forms (lignin as such, lignin-based polyols, depolymerized and fractionated lignins) will be presented. Face to the actual context, lignin is consolidating its role in the field of polymeric materials with some successful applications. The market is expected to benefit from major industrial manufacturers investments in R&D, such as in the development of improved technologies for lignin extraction and development of new products and applications. A crucial point is to achieve standardized raw lignins for industrial use.

Keyword Polyurethanes · Polyols · Lignin liquefaction · Oxypropylation of lignin · Rigid poliurthane foams

3.1 Overview of Strategies and Opportunities

There has been a long standing in industry saying that "One can make everything from lignin, except money" (Ragauskas et al. 2014). This situation leads to continuous research efforts to valorize lignin, and presently, different approaches to generate value-added products from this renewable raw material can be found (Cateto 2008; Duval and Lawoko 2014; Gandini 2011; Gandini and Lacerda 2015; Lange 2007; Matsushita 2015; Silva et al. 2009). The role of lignin as a renewable source, and the associated technical barriers for its utilization, was systematized, some years ago, in a report from the Department of Energy of the United States of America (Holladay et al. 2007). This report categorized the opportunities for lignin-derived

© Springer Nature Switzerland AG 2018
A. E. Rodrigues et al., *An Integrated Approach for Added-Value Products from Lignocellulosic Biorefineries*, https://doi.org/10.1007/978-3-319-99313-3_3

products in near-term (current uses and those viable within 3–10 years), medium-term (5 to perhaps 20 years), and long-term opportunities (more than 10 years and requiring significant knowledge investment and technology development). In fact, the complex macromolecular structure of lignin encompasses a wide range of possibilities with different high-value products able to be generated. Generally speaking, these products fall into three main categories: power/fuel, macromolecules, and aromatics (Luo and Abu-Omar 2017).

Current commercial uses of lignin take advantage of lignin's polymeric structure and polyelectrolyte properties. According to a recent market study, the global lignin market value was estimated in 2015 at $732.7 million, with Europe dominating the sharing (Grandviewreport 2017). The increasing demand of lignin for binders, adhesives, and concrete admixtures is expected to raise the market growth over the 2014–2025 period. In addition, the growing demand for dust control systems due to air pollution intensification and the successful use of lignin as concrete additive are likely to emerge as the major growth factors over the next years. Also, REACH (Registration, Evaluation, Authorization and Restriction of Chemicals) regulations, which aimed at restrict transportation of synthetically derived chemicals, are expected to push lignin market over the coming years.

The development of renewable lignin-based polymers is primarily driven by the need to reduce the dependency from fossil fuel derivatives raw material feedstocks. In the last decades, the exploitation of lignin-based monomers for polymer synthesis has attracted a major attention, owing to lignin's low cost, easy availability, renewability, and unique properties. Factors like petroleum market instability, increase global warming emissions, and the need to reduce energy demand drive the need to find alternatives to petrochemical-based sources. And so, the traditional raw materials are becoming partially substituted by biomass-based counterparts, namely, lignin-based alternatives, but still needing investment toward the development of economically feasible manufacturing processes. A strong global industrial and academic trend is to look for greener solutions in the field of polyols production, maintaining their properties and quality and thus impact in a series of derived polymeric materials (e.g., polyurethanes and polyesters). The market is expected to benefit from major industrial manufacturers investments in R&D, such as in the development of improved technologies for lignin extraction and in the development of new products and applications (Grandviewreport 2017).

Currently, the main lignin applications are dispersants, emulsifiers, binders, and sequestrants corresponding to nearly three-quarters of the available commercial lignin products. Generally, for these applications, lignin suffers little (most often sulfonation or hydroxymethylation) or no chemical modification; they represent relatively low value and limited volume growth applications. On the other hand, the use of lignin as a macromonomer in polymer synthesis requires fundamental understanding of lignin reactivity and its relation with raw material sources and isolation processes, to which is important to achieve standardized products for industry. This knowledge is crucial leading to the development of appropriate technologies for modification, control chemical reactivity, including copolymerization, and compatibility with other monomers and polymers (Holladay et al. 2007). Fortunately, it is becoming clear that lignin is a suitable material for the production of enhanced composites, carbon fibers, polyurethanes, polyethers, polyesters, and phenolic resins, with adding environmental benefits. Furthermore, other lignin-based products such as adsorbents, flocculants, adhesives, antioxidants, energy-storing films, and vehicles for drug delivery and gene therapy can be obtained (Norgren and Edlund 2014; Ten and Vermerris 2015).

The main difficulties associated with the development of lignin-containing polymeric materials can be described according to the following points: (i) heterogeneity of the commercially available technical lignins; (ii) variability of the hydroxyl groups in what concerns content, type, and proportion of aliphatic/aromatic; (iii) differences in molecular weight, which are dependent on the used biomass processing technology; (iv) poor solubility in organic solvents; and (v) limited accessibility to reactive sites for isocyanate reaction due to steric hindrance (Chauhan et al. 2014; Matsushita 2015; Upton and Kasko 2016).

In what concerns the use of lignin in polyurethane synthesis, two general approaches can be followed: (1) the direct use of lignin, i.e., without any preliminary chemical modification, alone, or in combination with other polyols, including bio-based polyols, or (2) the use of chemically modified lignin, by making hydroxyl functions more readily available, e.g., through esterification and etherification reactions. An overview of synthetic routes is presented schematically in Fig. 3.1.

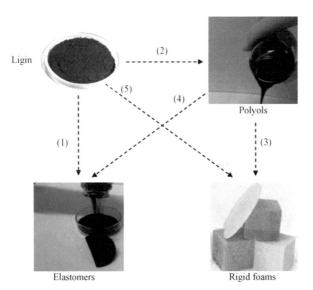

Fig. 3.1 Synthetic routes for lignin incorporation in polyurethane materials: (1) direct use, (2) synthesis of liquid polyol, (3) synthesis of rigid polyurethane foams using lignin-based polyols, (4) synthesis of polyurethane elastomers using lignin-based polyols, and (5) synthesis of rigid polyurethane foams using lignin as reactive filler. (Reprinted from Silva et al. (2009) from kraft lignin, 1276–1292, Copyright (2009), with permission from Elsevier)

The synthetic routes presented comprises the following strategies: (1) shows lignin direct use as macromonomer in polyurethane elastomers; (2) followed by (3) illustrates lignin liquefaction by chemical modification and subsequent use in rigid polyurethane (RPU) foams; (2) followed by (4) illustrates the preparation of an intermediate lignin-based polyol and its subsequent incorporation into polyurethane elastomers, and (5) corresponds to the direct use of lignin as a reactive filler in polyurethane foam chemical systems (Silva et al. 2009).

3.2 Lignin Use as Such

Due to the presence of both aliphatic and aromatic hydroxyls on its structure, which can act as reactive sites for isocyanate conversion, lignin can assume a direct role of a comonomer in polyurethane synthesis. However, the use of these strategies often result in rigid and brittle materials and give rise to the incorporation of modest lignin quantities. Lignin molecules are stiff polyhydroxyl macromolecules, thus generating highly crosslinked three-dimensional polyurethanes after reacting. Also, lignin alone can be either too unreactive or too insoluble, to ensure its homogeneous incorporation in the reactive mixture. To surpass these constraints, lignin is often combined with a polyol that can act as a solvent for lignin and provide flexibility to the final products (e.g., in the case of polyurethane elastomers).

In this context, this topic will cover two main polymeric applications where lignin is used directly: the use of lignin as reactive filler in polyurethane foams and the use of lignin as a comonomer in the synthesis of polyurethane elastomers. Lignin, combined with different polyols (petroleum- and natural-derived polyols), was tested. In a general way, foams with increased mechanical properties and thermal resistance were obtained. In a different perspective, lignin used as an additive to enhance biodegradability of thermoplastic polyurethanes will be also presented. Through this strategy, the filler (lignin) is dispersed into the polymeric matrix by compounding, giving rise to preferential sites for microorganism's attack, thus favoring biodegradation initiation and progression.

3.2.1 Reactive Filler in Polyurethane Foams.

Rigid polyurethane (RPU) foams are materials with low density, low thermal conductivity, and low moisture permeability properties, along with a high strength to weight ratio performance attributes. These properties have made RPU foams one of the most used polymeric foam materials on a global basis. Raw lignin, without any chemical modification, has been directly incorporated into RPU foam formulations to improve, or enhance, properties (Kai et al. 2016). In this context, several studies were performed in order to analyze the use of different lignins as reactive fillers in RPU foams. Also, several applications were envisaged, from the typical insulating materials to materials for dyes and crude oil sorption. In addition to RPU foams, some examples of lignin use in flexible polyurethane formulations will be also described.

Sodium lignosulfonate (SL) was studied in the synthesis of RPU foams combined with molasses (M) using different SL/M weight ratios (100/0, 80/20, 60/40, 40/60, 20/80, and 0/100; w/w). Foams, produced at various isocyanate/hydroxyl (NCO/OH) molar ratios, were evaluated concerning thermal properties by differential scanning calorimetry (DSC), thermogravimetric analysis (TGA), and thermal conductivity. Glass transition temperature (Tg) was dependent on the used NCO/OH molar ratio and varied between 80 and 120 °C, whereas thermal decomposition temperature (Td) was in the range 280–295 °C. Tg and Td were scarcely affected by the used SL/M ratio. Thermal conductivity of the produced foams was in the range from 0.030 to 0.040 Wm^{-1} K^{-1}, being dependent on the used SL/M ratio (Hatakeyama et al. 2008). In another study of the group (Hatakeyama et al. 2013), sodium lignosulfonate (NaLS)-based RPU foams were prepared using different kinds of glycols (ethylene glycol, diethylene glycol, triethylene glycol, or polyethylene glycol). Two types of industrial NaLS, acid-based and alkaline-based, were selected and mixed at different ratios with the glycols (0–16%, w/w). In terms of reactivity, it was observed that reaction time increased as the acid-based NaLS content in polyols increased and that apparent density, compression strength, and compression modulus of the RPU foams linearly increased with the reaction time. Glass transition temperature, measured by DSC, can be modulated in the temperature range of 310–390 K by changing polyol composition (acid-based/alkaline-based NaLS ratio and overall content and type of glycol).

Lignin derived from a bioethanol process was studied as a reactive filler to produce soy-based RPU foams. The chemical system comprised soybean oil-derived polyol, a petrochemical polyol (Jeffol A-630), and poly(diphenylmethane diisocyanate) (pMDI). The RPU foams were produced in free-rise mode using water as the blowing agent and lignin at contents of 0, 5, 10, and 15%. As the lignin content increased, the overall cell structure becomes more uniform, even cell walls became thinner and the number of damaged cells increased. The presence of lignin was also ascribed as influencing the cell nucleation process. Additionally, density increased as a result of lignin content increment, as well as mechanical properties (compressive strength, compressive modulus, flexural strength, and impact strength), up to 10%. For the assay with 15%, a decrease on mechanical properties was observed. Tg and storage modulus increased over the values observed for the reference foam (sample without lignin), and thermal stability increased with lignin content. Dynamic mechanical analysis (DMA) corroborated the increased mechanical properties and thermal stability of the lignin-based RPU foams (Luo et al. 2013).

Wheat straw lignin is being the focus of different works. These include the examples of the work of Paberza et al. (2014) and the work of Li et al. (2017b). In the work of Paberza et al. (2014), the production of RPU foams from tall oil polyol and an organosolv wheat straw lignin, added at contents ranging from 0% to 6.3% (w/w), gives rise to RPU foams with an apparent density of 45–60 kg/m^3. The use of lignin, combined with the bio-based polyol, increased the renewable content of the final foam up to 23.6% (w/w). The maximum value of compressive strength (0.35 MPa) was reached for RPU foams prepared with a lignin content of 1.2% (w/w). The produced materials were found adequate for thermal insulation applications. In the work of Li et al. (2017b), wheat straw lignin was used at contents between 0% and 2.5% (w/w, polyol basis) envisaging the production of oily solvents adsorption materials. The obtained results showed that physical properties improved with crosslinking density and the process of adsorption was in accordance with the quasi-second-order kinetic model.

Considering the last envisaged application, RPU foams for crude oil sorption, in the work of Santos et al. (2017), foams were prepared using lignin at contents of 0–20% (w/w), resulting in a decrease of hydrophobicity as the lignin content increase. Thus, all produced foams showed an improvement in the oil sorption capacity, e.g., the RPU foam prepared with 10% lignin (RPU10) showed an improvement of about 35.5% comparatively with the bank sample (RPU without lignin). Langmuir isotherms fitted well experimental data and predicted a maximum oil adsorption of 28.9 g/g in tests using the sample RPU10.

Organosolv and kraft lignins were used by Pan and Saddler (2013) to replace a petroleum-based polyol at contents ranging from 12% to 36% (w/w) for organosolv lignin and 9–28% (w/w) for kraft lignin. Lignin was able to chemically link to the polymer network, beyond being physically trapped in the final materials. The lignin-containing foams had comparable structure and strength up to 25–30% (w/w) for organosolv lignin-based foams and up to 19–23% (w/w) for kraft-based foams. The results indicated that organosolv lignins, which can be used at higher contents, give rise to materials with better performance, the fact associated with its superior miscibility. Chain extenders, such as butanediol, were also tested as comonomers, improving the strength of lignin-containing RPU foams.

Kraft lignins were also evaluated in the production of semirigid PU foams combined with bio-based polyols and biomass industrial wastes (Carriço et al. 2016). The foams were synthesized from physical mixtures comprising lignin, castor oil, and residual glycerol. Lignin content increase (10–40% (w/w)) resulted in density rise and thermal stability decrease. The foam synthesized with 17.5% lignin presented the best dimensional stability and thermal properties, together with higher cell homogeneity. This formulation was studied concerning the effect of castor oil content (relative to glycerol), NCO/OH molar ratio, and type of blowing agent. The increment on castor oil content, as well as the increment on the NCO/OH molar ratio, gives rise to foams with higher density and compressive strength, the fact associated with the strengthening of the crosslinking density. Comparatively with water (chemical blowing agent), physical agents (cyclopentane and n-pentane) originated foams with higher densities and lower compressive strengths. In conclusion, green PU foams with good properties and industrial interest were produced from residual wastes (glycerol and lignin) and castor oil, a bio-based polyol.

Pine needles, considered as an important and abundant bio-waste, were used to extract lignin that was further used in the synthesis of RPU foams to be employed as adsorbent for dyes (Kumari et al. 2016). The results demonstrated that the produced RPU foams were efficient in the removal of cationic dyes (e.g., malachite green from aqueous solutions) becoming less effective with the anionic ones. The adsorption kinetics and isotherms fitted well the pseudo-second-order model and Langmuir adsorption isotherm, respectively, with a maximum adsorption capacity of 80 mg/g. In addition, lignin-based RPU foams were reusable, having a cumulative adsorption capacity of 1.33 g/g after 20 regeneration cycles. In a different work, lignin-based RPU foams were reinforced with different weight ratios of pulp fiber (1, 2, and 5%). Morphological analysis showed the presence of inhomogeneous, irregular, and large-size cells, being significantly influenced by lignin and pulp fiber contents. Furthermore, this modification resulted in materials with decreased densities. As the lignin content increased, compressive strength decreased, and no beneficial effect derived from the adding of pulp fiber was observed. TGA results evidenced the formation of a high carbonaceous residue associated with lignin presence, while the incorporation of pulp fiber slightly improved the thermal stability of the lignin-based RPU foams.

Considering lignin use in flexible PU foam formulations, Jeong et al. (2013) studied the production of water-blowing flexible PU foams using softwood kraft lignin combined with polyethylene glycol of different molecular weights and 2,4-toluene diisocyanate (TDI). The contribution to the overall viscoelastic properties of the filler-like behavior surpassed lignin crosslinking effect; the resultant viscometric properties increased, as the lignin content increased. The produced materials were considered suitable for cushioning use. Moreover, the cyclic compressive tests evidenced a better shape recovery performance for the produced foams with high density. In a different work, Fernandes et al. (2014) studied the production of flexible polyurethane foams modified with lignin-based fillers for active footwear insoles. The objective was to improve properties such as resistance to fatigue and cushioning. With this purpose, the effect of using lignin as bio-based filler was tested with a flexible PU foam base formulation. The effect on maximum deceleration and energy return was studied. Foams were prepared at laboratorial scale by using the closed mold synthesis in order to control density. Lignin contents of 1 and 2% (w/w) were tested. Properties, namely, maximum deceleration and energy return, were measured according to the impact absorption drop test adopted by the Portuguese Footwear Technological Centre (CTCP). The obtained values were compared against the internal specifications also established by CTCP (casual shoes: maximum deceleration <150 m/s^2 and energy return $>34\%$). The results were compared with the base material alone or modified with 0.5 and 1% (w/w) of glycerol. Moreover, the effect of the sample thickness on the measured properties was inspected trying to approach the real insole dimensions. Among the tested samples, the use of lignin as filler (2%, w/w) gave quite interesting results, being the maximum deceleration 138.8 m/s^2 and the energy return 49.6% (used thickness of 16.5 mm). These results accomplished the specifications, indicating the suitability of the PU foam modified with 2% of lignin for the production of footwear insoles with improved cushioning properties.

3.2.2 Additive to Enhance Biodegradability

Nowadays, polymeric waste is becoming a serious environmental problem worldwide. In fact, since the development of the first polymeric materials, research efforts were concentrated on increasing their stability and extend lifetime. Given their durability, polymeric materials become attractive choices giving rise to their intensive use and consequent generation of waste at a global level. Taking into account that respect for the environment is a key issue in sustainable development, it became necessary to redefine strategies, processes, and products in order to preserve fossil resources and minimize pollution. It is thus essential that industries reduce energy consumption and intensify the use of raw materials derived from renewable natural sources. In this context, the development of strategies for increasing the biodegradability of polymeric materials in general, and polyurethanes in particular, at their end-of-life use, represents an important issue and has been the focus of several studies. Among the possible choices to enhance polymer's biodegradability, the use of natural-derived fillers, namely, lignin, starch, or cellulosic fibers, is a feasible alternative. Depending on the polymer type, these additives can be added during polymer synthesis, before or during processing stage. Through this strategy, the filler is dispersed into the polymeric matrix, thus acting as preferential sites for microorganism's attack, favoring biodegradation initiation and progression.

Concerning biodegradability, lignin is generally considered a recalcitrant part of plant cell walls, being mostly degraded by fungi via oxidative processes. The organisms predominantly responsible for lignin degradation are fungi, being the basidiomycetes the most effective degraders within this group. The ability to degrade lignin is thought to be associated with mycelial growth, which allows the transport of nutrients. The fungal degradation is mainly an extracellular process, due to the lignin insolubility. Fungi have two types of extracellular enzymatic systems: the hydrolytic system, which produces hydrolases that are responsible for polysaccharide degradation, and oxidative extracellular ligninolytic systems, which degrade lignin by opening phenyl rings. Laccases and peroxidase enzymes cause lignin degradation through low molecular weight free radicals such as OH, depolymerize the phenolic and non-phenolic groups, and mineralize the insoluble lignin (Datta et al. 2017; Sanchez 2009). Among the existing fungus, white-rotting fungi are the sole fungi in nature able to mineralize lignin, being *Phanerochaete chrysosporium* considered the model white-rotting fungi encompassing these skills (Camarero et al. 2014; Sanchez 2009; Zhu et al. 2017a). Beyond this, other fungi like the wood-rotting fungi have also the capability to produce ligninolytic oxidoreductases (laccases and different types of peroxidases), which are able to oxidize the phenylpropane lignin units (Liew et al. 2011; Suman et al. 2016). Despite all the aforementioned, some bacterial genre have also the ability to metabolize lignin. Soil bacteria actinomycetes have been reported to degrade lignin, both by solubilizing it and producing high molecular weight metabolites. Although the bacterial metabolism of lignin is not as complete as for fungal systems, bacteria action results in small-size aromatics that can be transported into the cell for aromatic catabolism, which is widespread in soil bacteria (Brown and Chang 2014). Due to the ability of bioconversion of lignin, microorganism's producing ligninolytic enzymes present also an increased capability to biodegrade several xenobiotic substances, including polymeric materials (Yadav and Yadav 2015). Nevertheless, potential applications using lignin-degrading organisms and their enzymes have become attractive since they can provide both environmental friendly technologies for pulp and paper industry and added value to lignin-based products that can be presented as more degradable materials (Amaral et al. 2012).

In the work of Ignat et al. (2011b), a series of lignin-PU blends were synthesized from biodegradable poly(ester-urethane) elastomers containing lactate segments and flax lignin through casting from dimethylformamide solutions. In order to evaluate the potential of fungal peroxidase and laccase on the produced blends, the obtained films were incubated with solutions of these enzymes. It was found that lignin addition improved the structural organization of PU and consequently the mechanical properties of blends. Moreover, FTIR analysis displayed that both, lignin and PU, were affected by the oxidative enzymatic treatment, with prevalence on film surface. PU degradation was mainly related with chain scission of polyester and lactate segments, being strongly influenced by lignin addition. The used enzyme type imparted a significant effect on degradation, and their increment influenced the thermomechanical behavior. In another study of the same group (Ignat et al. 2011a), the evaluation of flax lignin, added at contents of 4.2 and 9.3% (w/w), on the enzymatic degradation of a poly(ethylene adipate)-derived polyurethane, was evaluated. PU films, produced by casting, were incubated for 3 days at 30 °C in buffer solutions of fungal peroxidase and laccase extracted from *Aspergillus* sp. and then compared with the untreated films. Surface modifications due to the enzymatic attack were evaluated by FTIR and SEM, while the impact on bulk properties was assessed by tensile tests and TGA measurements. This study pointed out that lignin, used at amounts around 4–5%, increased both mechanical properties and propensity to biodegradability of common PUs based on MDI/polyester. Among the used enzymes, laccase showed the best degradation effects.

In the work of Fernandes et al. (2016a), the compounding of a polyester-based thermoplastic polyurethane (TPU) used in the footwear industry for outsoles production, with lignin, starch, and cellulose at content of 4% (w/w), was studied. The biodegradability was evaluated using agar plate tests with the fungi *Aspergillus niger* ATCC16404, the Gram-negative bacteria *Pseudomonas aeruginosa* ATCC9027, and an association of both (consortium), as well as soil tests conducted at 37 °C and 58 °C. The results obtained, both on the agar plate test and in soil tests, indicated a biodegradation-promoting effect associated with the use of the selected natural additives. Nevertheless, the best results were obtained with lignin, which give rise to a weight loss of 67% according to the biodegradation tests performed at 58 °C during 4 months (at this time the disintegration of the sample was observed). The FTIR analysis of these samples proved the effective degradation of the PU, in particular by the appearance of bands associated with the degradation of both urethane and ester groups, as well as the intensity reduction of the band at 713 cm^{-1} associated with the size reduction of the polymeric chain. Based on these results, the formulation TPU added with 4% lignin was selected for the production of soles' prototypes, and the physical-mechanical properties were evaluated. The results prove the suitability of using the lignin-modified TPUs for footwear outsoles' production.

3.2.3   Comonomer to Produce Elastomers

Regarding lignin utilization as a comonomer in the production of polyurethane (PU) elastomers, several studies were performed in order to evaluate the effect of lignin type and content, chemical system, formulation, and process conditions on material's properties. In this context, it is a current practice to use a tri-component chemical system (lignin, polyol, and isocyanate). Examples of polyols include poly(ethylene glycol) (PEG), poly(propylene glycol) (PPG), and polycaprolactone (PCL) of different molecular weights. Polyol will contribute to reduce the glass transition temperature and the crosslinking density, thus ensuring the obtainment of more flexible materials. Additionally, they can assume a solvent role enabling polymerization in bulk, thus avoiding the use of organic solvents.

The direct use of the tri-component system was extensively studied by Thring and co-workers (Thring et al. 1997, 2004; Vanderlaan and Thring 1998) who studied the effect of lignin content and NCO/OH molar ratio, lignin molecular weight, and polyol molecular weight. Additionally, Ni and Thring (2003) also studied the effect of using a catalyst (DBTDL). The work performed by Thring and co-workers used an organosolv lignin (Alcell from Repap) and showed that the use of high lignin contents (greater than 30%) and high NCO/OH molar ratios lead to the obtainment of hard and brittle materials. The crosslinking density of these materials was found to increase with the increase of lignin molecular weight. The inclusion of DBTDL allowed the utilization of higher NCO/OH molar ratios as well as higher lignin contents (50%) without leading to the formation of brittle materials. The results were explained in terms of catalyst ability to efficiently promote the reaction between the polyol's hydroxyl groups and isocyanates.

Hatakeyama's research group also performed a series of works using this strategy. The chemical system comprised of a kraft lignin, a polyether polyol, and polymeric MDI (Reimann et al. 1990; Yoshida et al. 1987, 1990). The effect of lignin content on crosslinking density and mechanical properties was evaluated at different NCO/OH molar ratios. Lignin was found to act as a crosslinking agent, and its contribution to the formation of the polymeric network was more efficient at low NCO/OH molar ratios. It was possible, by varying the lignin content, to produce polyurethanes with a wide range of mechanical properties, using low to intermediate NCO/OH molar ratios. This work was also extended to other lignin types, namely, solvolysis lignin and lignosulfonates (Hatakeyama 2002; Hirose et al. 1989). In these works, the contribution of lignin to the thermal properties of the resulting materials was also reported and found to retard thermal degradation of the resulting materials.

Gandini's research group contribution to this kind of approach includes the utilization of a kraft lignin alone or in combination with poly(ethylene oxide) (PEO) and hexamethylene diisocyanate (Cheradame et al. 1989). Other works of the group used a distinct approach to introduce the flexible segment (polyol), namely, by using oligoether isocyanates to combine with an organosolv lignin (Alcell from Repap) (Gandini et al. 2002; Montanari et al. 1996) and oxygen-organosolv lignins isolated from spent liquors after delignification of aspen and spruce in different acidic/organic solvente (Evtuguin et al. 1998). The resulting materials varied from thermoplastic to partly or fully crosslinked elastomers, depending on the functionality of the used isocyanates.

In the work of Cateto et al. (2011), the formation of lignin-based polyurethanes was monitored through Fourier-transform infrared spectroscopy working in attenuated total reflectance mode (FTIR-ATR). The used chemical system consisted of 4,4′-methylene-diphenylene diisocyanate (MDI), polycaprolactone (PCL) of three different average-molecular weights (400, 750, and 1000), and two lignin samples (Alcell and Indulin AT) incorporated at different weight contents (10%, 15%, 20%, and 25%) in the polyol mixture. The polymerization reaction was carried in bulk and without the presence of any catalyst. Results showed that isocyanate conversion decreased with the increase of both lignin content and PCL molecular weight. Moreover, the Indulin AT series leaded to lower isocyanate conversion comparatively with Alcell counterparts. The global second-order kinetic treatments showed to be dependent on lignin type and content as well as on PCL molecular weight, being more adequate for experiments with low lignin contents. The obtained results were systematized graphically in Fig. 3.2. In the network structure, Indulin AT generates larger and stiffer lignin "islands" comparatively to Alcell (Indulin AT has a higher molecular weight). Low molecular weight PCL are less efficient in creating three-dimensional polyurethane networks, effect highlighted for Alcell lignin (Alcell has a lower hydroxyl content). This was markedly observed for the sample synthesized from PCL400 with a lignin content of 25%, for which an apparently abnormal high sol fraction was achieved. On the opposite side, high molecular weight polyols increased the efficiency of the polymeric network formation. For very long PCL chains, this efficiency decreases mainly due to a dilution effect of the hydroxyl chain ends. This effect was highlighted when combined with Indulin AT where the volume of the stiff lignin islands hinders an effective network formation. This observation justifies the results obtained for the sample produced with PCL1000, for which also an apparently abnormal high sol fraction was achieved (Cateto 2008).

The present approach of using a three-component system (lignin, polyol, and isocyanate) continues to be explored in the recent years with some works focusing on the use of bio-based polyols putting in evidence the trend toward the increase of the bio-based content of the final materials. This is the case of the work of Tavares et al. (2016) where castor oil and also a modified form of castor oil (obtained through reaction with glycerol using different glycerol/castor oil weight ratios) was combined with a kraft lignin and MDI. The obtained results pointed out for an increase of the glass transition temperature and crosslinking density, together with

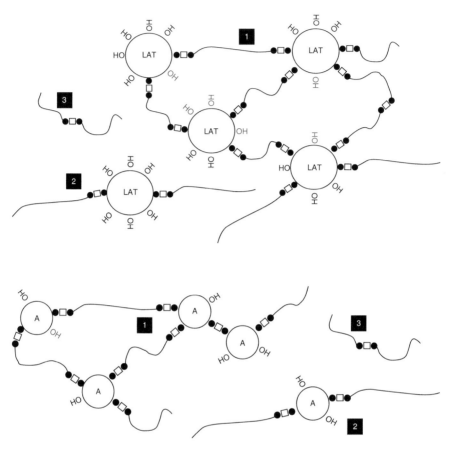

Fig. 3.2 Schematic representation of a lignin-based polyurethane network putting in evidence the structural differences between Alcell (A) and Indulin AT (IAT) (1, network; 2, isolated sequence containing lignin; and 3, isolated sequence without lignin). Possible entrapped hydroxyls are highlighted in red

an improved ultimate stress, for films prepared with modified castor oil added with 30% of lignin.

In the work of Avelino et al. (2017), polyurethanes using a system formed by a coconut shell-derived organosolv lignin, polyethylene glycol 400, and toluene diisocyanate (TDI) were synthesized through a solvent-free polymerization route. Properties such as porosity and crosslinking degree were determined having in view the evaluation of the lignin content (0–50%) and NCO/OH molar ratio (1.0–1.5). From the thermal analysis results, it was concluded that the lignin-based PU presented higher thermal and thermo-oxidative stability, when compared with the base PU (sample without lignin). Glass transition temperature increased with lignin content increase, compatible with the increase of the crosslinking density due to effective lignin incorporation. High lignin contents produced PUs with poor mechanical properties, while intermediate lignin contents produced materials with interesting mechanical properties.

A different approach, but in line with the one tested by Gandini's group (the use of oligoether isocyanates), was recently studied by Li et al. (2017a), namely, by the use of poly(propylene glycol) tolylene 2,4-diisocyanate terminated (PPGTDI) as soft domains and lignin as the hard segment. Two lignins of different molecular weights (600 and 3600) were tested at contents ranging from 5% to 40%. With lignin content increase, PU thermal stability was improved, and Tg shifts to higher temperatures, especially for the PUs produced with the low molecular weight lignin. Furthermore, these samples displayed improved mechanical properties. The sample comprising 40% of the low molecular weight lignin presented Young's modulus, tensile strength, and strain at break of 176.4 MPa, 33.0 MPa, and 1394%, respectively. This can be related with the better dispersion of low molecular weight lignin in the reactive mixture that favors lignin reaction and incorporation in the PU network, as verified by DSC, SEM, and TEM studies. In conclusion, this study highlighted the potential application of unmodified lignin in the high-performance PU production. The used strategy resulted in final PU samples with high lignin contents (40%, PU basis).

3.3 Lignin Use After Chemical Modification

Lignin chemical modifications can be categorized into three main classes: (i) lignin functionalization/derivatization having in view the introduction of new chemical active sites, thus enhancing lignin reactivity/compatibility in polymeric systems; (ii) lignin chemical modification to produce liquid polyols, a comonomer for polyurethane synthesis; and (iii) lignin fragmentation/depolymerization procedures, which increase lignin compatibility with other comonomers and solvents (Lora and Glasser 2002; Panesar et al. 2013).

In this context, current practices to functionalize lignin, namely, processes involving etherification (e.g., lignin methylation), esterification (e.g., lignin acetylation), and reaction with isocyanates, will be briefly reviewed. Then the topic of the production of liquid polyols by liquefaction procedures using polyhydric alcohols via solvolytic reactions (processes which can be also considered as depolymerization processes) and by oxypropylation (processes that preserve the macromolecular characteristics of lignin) will be addressed in more detail due to the interest for polyurethane chemistry. Following the current interest on fragmentation/depolymerization processes, which include other strategies over the cited liquefaction processes, this topic will be treated in an independent item.

Lignin Functionalization/Derivatization
Lignin functionalization/derivatization aims to introduce new chemical active sites within its structure in order to render it more reactive/compatible enhancing its use in the synthesis of polymeric materials. Lignin reactivity is essentially based on two

distinct structural features: (i) the presence of free ortho positions in the phenolic ring (guaiacyl-type lignins from softwood can comprise reactive sites on C5 positions of the phenolic ring and lignins containing coumaryl-type units (e.g., some lignins from annual plants) can present reactive sites on C3 positions) and (ii) the presence of multiple hydroxyl functions (OH), which can be subjected to various chemical modifications. Depending on the envisaged chemistry for OH's functionalization, it is needed to differentiate between the aliphatic (primary or secondary) and the phenolic ones (condensed or non-condensed) (Duval and Lawoko 2014; Mahmood et al. 2016b). Among the available functionalization/derivatization procedures, etherification, esterification, and reaction with isocyanates are the ones most explored. These chemical modifications involve hydroxyl groups.

Methylation can be viewed as a method mainly to improve lignin compatibility with nonpolar polymeric matrices (Duval and Lawoko 2014). Also, it can be used to reduce lignin's hydroxyl functionality. Lignin methylation was studied by Sadeghifar et al. (2012) who evaluated and compared two series of methylated softwood kraft lignins synthesized using different methylation pathways. The obtained data showed that, under the tested conditions, dimethyl sulfate in aqueous alkaline medium (NaOH) selectively converts phenolic hydroxyl groups of lignin into methylated derivatives (side reactions were considered negligible). On the other hand, methyl iodide in the presence of an excess of potassium carbonate in N,N-dimethylformamide was found to be ineffective and nonselective. In another study (Cui et al. 2013), the effect of using methylation (through dimethyl sulfate in alkaline medium) in the thermal properties of a softwood kraft lignin was inspected. Softwood kraft lignin is generally susceptible to thermally induced reactions which cause modifications in the molecular structure, namely, by the formation of irreversible crosslinking. The fully methylated lignin presented an increased thermal stability, together with a low Tg, minimizing these undesirable side effects. These results were also corroborated in the work of Sen et al. (2015) who studied methylation using dimethyl carbonate (DMC) in the presence of sodium hydroxide or cesium carbonate as bases, under mild reaction conditions. Moreover, the results evidenced the preferential methylation of the phenolic hydroxyls, while for the aliphatic ones, a reduction was observed. Methylated samples presented higher thermal stability, avoiding induced crosslinking.

Regarding esterification, Chen et al. (2014) studied lignin esterification (kraft lignins were used) with maleic anhydride, succinic anhydride, and phthalic anhydride, being phthalic anhydride the reactant conducting to the highest weight gain. Spectroscopic analysis revealed a decrease in hydroxyl content and an increase in carbonyl content, which was related with the effective esterification. As a consequence, the hydrophilic properties of the modified lignin decreased. Regarding the thermal stability, maleic and succinic anhydrides gave rise to modified lignins with increased thermal stability. The ones modified with phthalic anhydride showed a rapid degradation at low temperature. In the work of Xiao et al. (2001), esterification with succinic anhydride was applied to several lignin samples (e.g., poplar, maize stem, barley, wheat, and rye straw lignins). All the modified lignins showed an increased thermal stability when compared with the corresponding unmodified lignins. Also in this context of lignin esterification, Cachet et al. (2014) studied the esterification of an organosolv lignin (Biolignin™) with acetic anhydride (acetylation) in supercritical carbon dioxide (scCO2). Comparing the supercritical conditions with the conventional reactions in solution, it was noticed that the supercritical medium allowed an improved acetylation yield, being also observed a decrease on the glass transition temperature of the modified lignin. Based on the overall results, it is possible to change the basic properties of lignins through reaction with anhydrides, a useful strategy for the production of high-performance composites.

Modification of lignin through reaction with isocyanates was the subject of several studies. In the work of Chauhan et al. (2014), soda lignin was functionalized with 4,4′-diphenyl methane diisocyanate (MDI) and thereafter reacted with PEG400. Gómez-Fernández et al. (2017) used a similar strategy by functionalizing a commercial kraft lignin through the reaction with isophorone diisocyanate (IPDI). The results revealed that almost half of the lignin's hydroxyl groups were functionalized. Thereafter, the modified lignin was used to produce flexible polyurethane foams. Two series of polyurethane foams, using modified and unmodified lignin (3, 5 and 10%), were prepared. The foams prepared with the modified lignin give rise to final materials with increased flexibility (higher elastic modulus and higher ability to absorb energy was observed), when compared with the ones using unmodified lignin. This was associated with a more effective incorporation of lignin through chemical reaction in the case of using the modified lignins.

In a recent work, Borrero-López et al. (2017b) tested the modification of an alkali lignin with different diisocyanates (hexamethylene diisocyanate (HDI), isophorone diisocyanate (IPDI), toluene diisocyanate (TDI), and 4,4′-diphenyl methane diisocyanate (MDI)). The production of polyurethanes involved a two-step procedure; first lignin was functionalized with the chosen diisocyanate and thereafter reacted with castor oil resulting in materials with gel-like type properties. Gel's (produced with 30% thickener) rheological properties were evaluated, being the HDI-functionalized lignin-based gels the ones presenting the higher gel-like type behavior. TDI-, IPDI-, and especially MDI-functionalized lignin-based gels showed weak gel-like type properties or even a liquid-like behavior. These features were related with the respective isocyanate chemical structures, which led to higher steric hindrance and thus to the diminishing of urethane linkages formation. This study was the base for the proposal of a lignin-based thickener for lubricant greases, application that was studied in a second work of the group (Borrero-López et al. 2017a). The performed tests, typical of the lubricant industry (penetration and tribological assays), showed that linear viscoelasticity functions were affected by the used lignin/diisocyanate ratio and thickener content, while thermo-rheological analysis revealed a softening temperature around 105 °C. Additionally, when compared with traditional lubricant greases, the produced gels presented lower friction coefficients.

Lignin modification by esterification and by reaction with isocyanates was compared in the work of Zhang and co-workers (2015). The aim of the work was to study the compatibility improvement of lignin fillers subjected to chemical modification with a PU matrix. Lignin modification was done following two strategies:

esterification with butyric anhydride and chemical modification with octadecyl iso-cyanate (urethane lignin). The modified lignins were incorporated in a vegetable oil-based polyurethane to produce composites with high bio-content. Between the two tested lignin fillers, urethane lignin showed better compatibility with the selected PU matrix, comparatively with the butyric anhydride modified lignin. It was observed that the incorporation of a high urethane lignin content (up to 30% (w/w)) increased the Young's modulus and lowered the onset of thermal degradation. However, urethane lignin incorporation did not affect significantly the glass transition temperature; nevertheless, an increase in the dielectric constant was identified, comparatively with the neat PU (PU without additives).

Beyond the chemical modifications aforementioned, lignin functionalization with phosphorus was also tested aiming at improve lignin compatibility with reactive PU systems and to increase lignin's flame-retardant properties. In this context, Xing et al. (2013) synthesized halogen-free flame-retardant RPU foams, where the base polyether polyol was partially substituted by a polyol prepared by a phosphorous-modified lignin added with encapsulated ammonium polyphosphate at contents of 10, 20, and 30% (w/w). Reference foams with the base polyol and with the defined polyol mixtures, without adding encapsulated ammonium polyphosphate, were also produced. The use of the phosphorous-modified lignin-based polyol and the encapsulated ammonium polyphosphate gives rise to RPU foams with improved thermal stability, flame-retardant, and mechanical properties. Also, it was observed that when the lignin-based polyol was used, the heat release rate decreases, and the combustion process is slowed down. In a similar work (Zhu et al. 2014), lignin grafted with phosphorus-nitrogen-containing groups, obtained via a liquefaction-esterification-salification process, was used to prepare lignin-based phosphate melamine compounds with free hydroxyl groups. This modified lignin was used to replace part of a base polyol to produce lignin-modified-polyurethane foams with flame-retardant properties. Comparatively with the base materials, the obtained foams showed improved compression strength, thermal stability, char formation, and self-extinguishment properties. The obtained features resulted in the improvement of the flame-retardant properties of the foams both on gas and condensed phases.

3.3.1 Overview of Lignin Liquefaction Processes

Considering lignin chemical modification having in view the production of liquid polyols, currently there are two major technologies: liquefaction and oxypropylation. The produced polyols are rich in hydroxyl groups and can be used for PU synthesis without further modification (D'Souza et al. 2017; Li et al. 2015). Liquefaction is a process in which a substrate is converted into smaller molecules by polyhydric alcohols via solvolytic reactions. Liquefaction can be applied to the whole biomass or their components, like lignin. According to several studies, the liquefaction process is usually conducted at temperatures in the range 150–250 °C under atmospheric pressure by using polyhydric alcohols, such as polyethylene glycol (PEG) and glycerol, as liquefaction solvents. Liquefaction can be either acid- or base-catalyzed, being however the acid catalyzed the most used process (Balat 2008). During lignin liquefaction, solvolysis and depolymerization phenomena occur simultaneously. Lignin is decomposed, depending on the used solvent through hydrolysis, phenolysis, alcoholysis, and glycolysis reactions, which cause biopolymers to cleave into smaller fragments. After this stage, the re-polymerization and condensation reactions of lignin fragments can occur increasing the final residue in the reaction medium (Silva et al. 2017).

Liquefaction can be performed with organic solvents using conventional heating systems or emerging microwave heating sources; this later conducting to a faster liquefaction process. Polyols obtained through liquefaction commonly contain solid residues in variable amounts depending on the process efficiency. Thus, the optimization of liquefaction parameters is usually conducted to obtain polyols with low residue content, which will be more suitable for the production of polyurethanes. Factors such as the feedstock characteristics, liquefaction solvent, catalyst, and liquefaction temperature/time have a significant effect on lignin liquefaction efficiency (Li et al. 2015). Table 3.1 presents a survey of examples dealing with the liquefaction of lignins coming from various processes (e.g., kraft, milled wood lignin, soda, and organosolv) and bio-based feedstocks (e.g., corncob residues, olive tree pruning, aspen wood, sugarcane bagasse, and cornstalk residues) and using different processing conditions.

Most of the liquefaction processes use sulfuric acid as catalyst at contents ranging from 1.0 to 3.0 (w/w), even some works report the use of much higher contents (10.0–20.0), as is the case of Sequeiros et al. (2013). Sulfuric acid and toluene sulfonic acid monohydrate (PTSA) catalysts were used by Jasiukaityte-Grojzdek et al. (2012) with beech (*Fagus Sylvatica*) milled wood lignin as a model substrate. In this work, sulfuric acid conducted to an increased formation of condensed structures when compared with PTSA-catalyzed liquefaction. Furthermore, PTSA favored lignin degradation and functionalization with ethylene glycol (used solvent). A gradual incorporation of the ethylene glycol into the lignin structure was observed, which give rise to a condensed lignin-based product with predominant aromatic hydroxyl groups. The presence of lignin monomers was also confirmed. Other alternatives of catalysts are ionic liquids (IL). In this context, 1-(4-sulfobutyl)-3-methylimidazoliumhydrosulfate was used in the direct liquefaction of sugarcane bagasse using an ethanol/water mixture as solvent (Long et al. 2015). The results demonstrated an excellent catalytic performance for the chosen IL. When using the optimized conditions, the degree of liquefaction surpassed 65%, yielding 13.5% of useful aromatic fine chemicals (e.g., phenol, 4-ethylphenol, and guaiacol).

As an alternative to conventional heating, microwave liquefaction is being progressively studied corroborating the advantages in what concerns the reduced reaction times. Microwave liquefaction of a kraft lignin was performed with a PEG400/glycerol mixture using sulfuric acid as catalyst (Silva et al. 2017). The work was focused on studying the effect of the lignin/solvent and catalyst/solvent ratios and reaction time on yield and hydroxyl index of the obtained polyols. The optimal

Table 3.1 Examples of lignin liquefaction processes and used conditions

Lignin	Proc. (Heat)	Solvent mixture (w:w)	L/S ratio (w:v)	T (°C) t (min)	Cat. (%, w/w)[1]	References
Milled wood (*Fagus sylvatica*)	G (C)	Ethylene glycol (100)	1:5	150 240	Sulfuric acid (3) p-toluene sulfonic acid monohydrate (3)	Jasiukaitytė-Grojzdek et al. (2012)
Kraft (*Eucalyptus* spp.)	G (M)	PEG400:glycerol (80:20)	2:10 1.5:10 2.5:10	160 5–30	Sulfuric acid (3–6)	Silva et al. (2017)
Alkaline (corncob)	G (M)	PEG400:glycerol (80:20)	1:5	160 5–30	Sulfuric acid (1.5)	Xue et al. (2015)
Enzymatic hydrolysis (cornstalk)	G (C)	PEG400:glycerol (80:20)	1:4 1:5 1:4.5 1:5.5 1:6	130–170 60–180	Sulfuric acid (10–20)	Jin et al. (2011)
Organosolv (olive tree pruning)	G (M)	PEG400:glycerol (80:20)	1.5:8.5	130–180 5–15	Sulfuric acid (1–3)	Sequeiros et al. (2013)
Hydrolysis (aspen wood)	H (C)	Ethanol:water (50:50)	2:10	150–300 60	n.u.	Mahmood et al. (2016a)
Organosolv (sugarcane bagasse)	H (C)	Ethanol:water (40:10)	0.5:50	250 15	1-(4-sulfobutyl)-3-methyl imidazolium hydrosulfate (1 mmol)	Long et al. (2015)
Soda (non-wood biomass)	G (M)	Glycerol:glycerin polyglycidyl ether (2:10, 3:10, 4:10; 5:10, 5:15)	1:12 1:13 1:14 1:15 1:20	140 2	n.u.	Bernardini et al. (2015)
Hydrolysis (residue from cornstalk)	A (C)	Furfuryl alcohol 100	1:3 1:4 1:5	120–170 15–120	n.u.	Li et al. (2013)
	G (C)	PEG400:glycerol (80:20)				
Kraft Organosolv Calcium Lignosulfonate	G (C)	Crude glycerol (100)	9:1	160 90	Sulfuric acid (n.a.)	Muller et al. (2017)

Lignin (L); solvent (S); temperature (T); time (t); process (Proc.), glycolysis (G); hydrolysis (H); alcoholysis (A); heat source (Heat), conventional (C); microwave (M); catalyst (Cat), solvent weight basis; n.u., not used; n.a., not available

liquefaction conditions, obtained through response surface methodology (RSM), were 20% of lignin, 3% of catalyst at 5 min giving rise to a liquefaction yield of 95.27%, and a hydroxyl index of 537.95 mg KOH/g. In another study (Xue et al. 2015), lignin-based polyols were produced by microwave heating at different times (5–30 min) using PEG400/glycerol (solvent) and sulfuric acid (catalyst) at 140 °C. The heating time of 5 min (optimal conditions) originated a high liquefaction yield (97.5%) and polyol with a hydroxyl number of 8.628 mmol/g, suitable for polyurethane foams synthesis. This was validated by producing PU foams with increased NCO/OH molar ratios (0.6–1.0), which resulted in mechanical properties improvement. Also, Sequeiros and co-workers (2013) evaluated the liquefaction of an organosolv lignin recovered from olive tree pruning, under microwave heating with PEG400/glycerol and sulfuric acid. The optimal conditions obtained corresponded to a liquefaction yield of 99%, obtained in 5 min at 155 °C, using 1% of sulfuric acid. The obtained polyol presented a hydroxyl index of 811.8 mg KOH/g. Microwave liquefaction was also applied to a soda lignin in glycerol/glycerin polyglycidyl ether medium (Bernardini et al. 2015). The obtained polyols were tested in the synthesis of flexible PU foams by partially replacing an oil-based polyol. The obtained foam's properties were similar to that of conventional materials used in furniture, car seats, and couches. According to this strategy, lignin polyol was incorporated in the PU foam at a content of around 12% (w/w).

Other studies comprise the use of different liquefaction solvents, such as monoethanolamine (Renata and Celeghini 2001), water/ethanol mixtures (Mahmood et al. 2016a), furfuryl alcohol (Li et al. 2013), and crude glycerol (Muller et al. 2017). According to this last work, where different technical lignins were studied (organosolv, kraft, and calcium lignosulfonate), polyols with adequate hydroxyl indexes for PU foams synthesis proposes were obtained (435, 515, and 529 mg KOH/g, respectively). Based on the obtained results, the authors concluded that crude glycerol can replace petroleum-derived liquefaction solvents.

3.3.2 Oxypropylation as a Viable Route to Produce Liquid Polyols

According to Gandini and Lacerda (2015), oxypropylation is, perhaps, the most interesting approach to use lignin in the field of materials synthesis. Oxypropylation is a very efficient process that converts quantitatively lignin proceeding from various industrial origins into viscous polyols (macromonomers for PU synthesis, namely, RPU foams). In fact, the preparation of low-cost polyols from abundant and renewable biomass resources is being an important topic of research, highlighted by the current industrial interest, namely, for polyurethane synthesis. In this context, simple sugars and other polyols such as glycerol are commonly oxypropylated resulting in some of the available commercial products. In what concerns lignin, oxypropylation has been recognized as a viable and promising approach to overcome the technical limitations and constrains imposed by its macromolecular nature, which limits its direct use. By means of oxypropylation, the hydroxyl groups, in particular the phenolic ones entrapped inside the molecule and of difficult access, are liberated from steric and/or electronic constraints becoming more accessible for reaction. At the same time, the solid lignin becomes a liquid polyol, thanks to the introduction of multiple ether moieties (Cateto 2008).

In a general way, oxypropylation reaction starts with hydroxyl groups activation by the catalyst, i.e., formation of an alcoholate group (RO-). Then, the formed alcoholate group attacks the oxiranic ring of a propylene oxide (PO) molecule yielding another alcoholate group, after the insertion of a PO moiety. This chain-extension reaction occurs up to total PO consumption. This attack is made preferentially at the α carbon of the oxiranic ring due to the low steric hindrance of this atom. Consequently, the hydroxyl groups will be predominantly secondary hydroxyls (94–96%) and the microstructure of the chain predominantly head-to-tail type (H-T) (90%). Although less probable, the attack of the alcoholate group can also occur at the β position of the oxiranic ring. This situation, considered to be abnormal, leads to the obtaining of primary hydroxyl groups (4–6%). The described reactions are in fact a repeated second-order nucleophilic substitution (SN-2 type) that attacks a strongly nucleophilic alcoholate group on the carbon atoms of the oxiranic ring (Cateto 2008).

Oxypropylation reactions are normally performed using a basic catalyst such as potassium hydroxide (KOH). Acid catalysts can also be used although they lead to the formation of substantial amounts of by-products (cyclic ethers). Other catalysts, such as polyphosphazenium, aluminum tetraphenylporphine, and cesium hydroxide, are also excellent to catalyze hydroxyl groups alkoxylation. Nevertheless, they are very expensive and therefore rarely used (Mihail 2005). Oxypropylation is always accompanied by the occurrence of two secondary reactions, PO homopolymerisation and isomerization. The homopolymerisation of PO takes places in the presence of residual water. This situation leads to the formation of OH species (e.g., from aqueous KOH), which can activate PO conducting to the formation of polyether diols (e.g., poly(propylene glycol)). PO isomerization results from the removal of a hydrogen atom of the PO methyl group leading to the formation of an allylate and finally to the initialization of PO anionic polymerization. In this case, polyether monols with a terminal double bond are obtained as side products (Cateto 2008). This transfer reaction is favored when high temperatures are used, and to a much less extent, at high catalyst concentrations (Mihail 2005). In lignin oxypropylation reaction, the formed side reaction products, mainly poly(propylene oxide) PPO oligomers, are normally left in the final mixture since they constitute a very useful bifunctional comonomer, decreasing viscosity and the glass transition temperature (Belgacem and Gandini 2008). Figure 3.3 represents schematically lignin oxypropylation process and the composition of the obtained polyol.

A survey of some representative works on the direct oxypropylation of biomass residues and isolated lignin, mainly to produce liquid polyols, are presented in

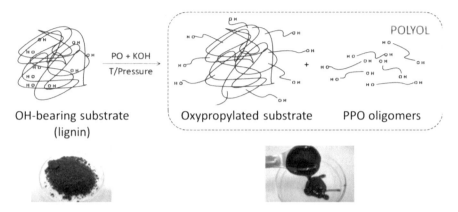

Fig. 3.3 Schematic representation of lignin oxypropylation reaction putting in evidence the final composition of the polyol (mixture of oxypropylated lignin and PPO oligomers)

Table 3.2. Apart from different types of lignin, examples include sugar beet pulp (Pavier and Gandini 2000), cork (Evtiouguina et al. 2002; Evtiouguina et al. 2000), olive stone (Matos et al. 2010), date seed (Briones et al. 2011), rapeseed cake (Serrano et al. 2010), and gambier tannin (Arbenz and Avérous 2015).

Lignin oxypropylation has been the subject of some pioneering work being worth to mention the one of Glasser (Glasser et al. 1983; Wu and Glasser 1984) and Gandini (Gandini et al. 2002; Nadji et al. 2005) research groups. Different lignins were oxypropylated, with or without catalyst, in bulk or solution medium, and using various lignin/propylene oxide (L/PO, w/v) ratios. Several catalysts (triethylamine (TEA), pyridine, diazabicyclooctane (DABCO), sodium hydroxide (NaOH), and KOH) have been tested at various contents. Among them, alkali metal hydroxides (KOH and NaOH), with particular relevance to KOH, were found the most efficient ones. Moreover, bulk oxypropylation showed to be advantageous; it gives rise to faster reactions and requires no solvent, which bring green connotations by minimizing the generation of volatile organic compounds (VOC) (Belgacem and Gandini 2008). Particularly, in the work of Nadji and co-workers (Nadji et al. 2005), a systematic study using different lignins (lignin from Alfa (*Stipa tenacissima*) (SL), organosolv lignin from hardwoods (OL), kraft lignin (KL) from softwood, and oxidized organosolv lignin (OOL)) was conducted. Oxypropylation was carried in batch mode (set-point temperature of 180 °C) using KOH as catalyst. The results indicated that lignins with a low molecular weight and high purity (hardwood organosolv lignins and non-wood lignins from soda pulping) reacted more readily with PO than the ones with high molecular weight and rich in carbonyl and carboxyl groups. The hydroxyl index of the obtained polyols was comprised between 100 and 200 mg KOH/g, making them suitable to be used in RPU foam formulations.

Another systematic work was the one conducted by Cateto and co-workers (2009) where an optimization study performed with Alcell, chosen as a model lignin, was performed using L/PO ratios (w/v) of 10/90, 20/80, 30/70, and 40/60 and

Table 3.2 Survey of oxypropylation conditions used with selected biomass residues and isolated lignins

Biomass or lignin type	B/PO (w/v) Cat. (%, w/w)	T_{set} (°C) T_{max} (°C)	$P_{max}/$ (10^5 Pa) t_{total}(min)	UR (w/w, %)	References
Sugar beet pulp	50/50–10/90 1–30	140 170	8 240	0–50	Pavier and Gandini (2000)
Cork	40/60–20/80 5 or 20	100–145 170–260	8–15 n.a.	0.5–5	Evtiouguina et al. (2000)
Cork	10/94 5 or 10	120–140 190–230	10–15 60–240	1–5	Evtiouguina et al. (2002)
Olive stone	30/70 n.a.	200 196	9 25	0	Matos et al. (2010)
Date seed	10/33–10/50 n.a.	160 260	27 n.a	7	Briones et al. (2011)
Rapeseed cake	10/23 n.a.	160 201	21 45	0	Serrano et al. (2010)
Gambier tannin	40/60–10/90 2.5–10	150 151–261	12.9–17.1 57–407	n.a	Arbenz and Avérous (2015)
Organosolv L (Straw)	15–40/ (140 g) 0.7–1.9	160–176 225–242	21.7–30.0 n.a	0.1–7.5	Arshanitsa et al. (2015)
Soda L (*Stipa tenacissima*) Organosolv L (hardwood) Kraft L (softwood) Oxidized lignin organosolv (*Picea excelsa*)	50/50–10/90 5–10	180 140–195	6.5–18 20–900	1–15	Nadji et al. (2005)
Kraft L (softwood) Soda L (non-wood) Organosolv lignin (hardwood)	40/60–10/90 2–5	160 170–280	15–25 35–110	0	Cateto et al. (2009)
Soda L (*Stipa tenacíssima*)	40/60–20/80 3–5	120–140 218–330	22–40 4–10	n.a	Berrima et al. (2016)
Kraft L (softwood)	10/40 5	150 285	17.5 9	n.a	Li and Ragauskas (2012b)
Sodium lignosulfonate (*Pinus taeda*)	40/60–25/75 5	150 240	20 48	n.a	Oliveira et al. (2015)

Lignin (L); biomass (B); propylene oxide (PO); KOH (Cat.); set-point temperature (T_{set}); maximum reached temperature (T_{max}); maximum reached pressure (P_{max}); total reaction time (t_{total}); unreacted biomass (UR) biomass-based; n.a., not available

catalyst contents in the range 2–5% (w/w, biomass-based). The reaction was carried out in bulk (set-point temperature of 160 °C) with KOH as catalyst. The effect of the selected variables on hydroxyl index, viscosity, homopolymer content, and molecular weight was analyzed. It was observed that, under conditions favoring hydroxyl activation (high catalyst content or high L/PO ratio), formation of short grafts occurs (lower hydrodynamic volume was detected by SEC, and high viscous polyols were obtained). In opposition, if hydroxyl activation is not favored (low catalyst content or low L/PO ratio), formation of longer graft chains takes place (higher hydrodynamic volume is detected by SEC, and low viscous polyols were obtained). The optimal oxypropylation conditions were established according to the requirements needed for the production of RPU foams (hydroxyl index between 300 and 800 and viscosity below 300 Pa.s). Based on these achievements, polyols based on four selected lignins (Alcell, Indulin AT, Curan 27-11P, and Sarkanda) were produced and used in the preparation of RPU foams.

Considering the main application of the lignin-base polyols, RPU foam synthesis, different types of lignins have been oxypropylated. Soda lignin precipitated from black liquor of *Stipa tenacissima* L. cooking (L/PO ratios of 20/80, 30/70, and 40/60, KOH (3–5%)) gives rise to polyols with viscosity and hydroxyl index ranging from 0.48 to 4.20 Pa.s and 150–375 mg KOH/g, respectively (Berrima et al. 2016). Straw lignin (Biolignin™) oxypropylation originated polyols with adequate properties, particularly if 15–30% of lignin was used in the initial reaction formulation. Further increase in lignin content lead to the formation of a non-liquefied fraction and high viscosity polyols (Arshanitsa et al. 2015). Kraft pine lignin and pine wood chips organosolv lignin were oxypropylated using a L/PO ratio (w/v) of 10/40 and KOH at a content of 1% (biomass basis) resulting in polyols fulfilling the required hydroxyl index (between 300 and 800 mg KOH/g)(Li and Ragauskas 2012a, b). In another study using sodium lignosulfonates, a polyol with a viscosity of 9.59 Pa s and a hydroxyl index of 574 mg KOH/g was produce being considered adequate for lignopolyurethanic composite preparation (Oliveira et al. 2015).

3.3.3 Screening of Opportunities for Oxypropylated Lignin

The use of lignin-based polyols in polymeric materials is mainly focused on the synthesis of RPU foams. This preferential choice is associated with the high functionality of these products. Even limited, studies envisaging other applications can be found in literature. In this context, Glasser and co-workers performed a series of studies that include the preparation of polyurethane films, coatings, epoxy resins, polymer blends (e.g., with poly(methyl methacrylate), poly(vinyl alcohol), and hydroxypropyl cellulose), and composite materials. This work remains as a relevant contribution in the field of lignin-based polymeric materials (Ciemniecki and Glasser 1988; Oliveira and Glasser 1994; Glasser et al. 1984, 1990, 1991; Hofmann and Glasser 1993; Oliveira and Glasser 1994; Saraf et al. 1985). The scenario did not change significantly in the present days, where only few works using oxypro-pylated lignins, or oxypropylated lignocellulosic biomass, to produce materials other than RPU foams, can be found. These include the modification of oxypro-pylated lignocellulosic biomass (Barbosa et al. 2013) and composite materials (Oliveira et al. 2015).

In the field of composite materials, lignin-based polyols prepared from oxypro-pylated sodium lignosulfonates were blended with castor oil and reacted with MDI to produce lignin-based composite matrices, which were reinforced with sisal fibers (Oliveira et al. 2015). It was observed that the adding of sisal fibers improved the impact strength of the produced lignin-based composite matrices. The produced composites resulted in materials with a wide range of properties.

To surpass the constraints associated with the high functionality of the oxypro-pylated lignocellulosic biomass, strategies for its decrease have been studied by reacting the oxypropylated materials with mixtures containing phenyl isocyanate (PI, a monofunctional isocyanate) and toluene diisocyanate (TDI, a bifunctional isocyanate) at PI/TDI molar ratios of 100/0, 80/20, 50/50, 20/80, and 0/100 (Barbosa et al. 2013). The objective was to modulate properties, thus opening new avenues for the exploitation of the oxypropylated products. The resulting materials showed that as the PI/TDI molar ratio decreased, Tg shifted toward higher values, observa-tion compatible with a crosslink density increase. In fact, the produced polyure-thanes changed from a highly viscous liquid (PI/TDI = 100/0 and 80/20) to stiff solids (PI/TDI = 50/50, 20/80, and 0/1 0/0). In conclusion, the strategy used in this work demonstrated a way to modulate the final properties of the oxypropylated products, pointing out the potential to be used in applications over RPU foams.

3.3.4   Production of Rigid Polyurethane Foams

Rigid foams' market evolution, with an expected growth by 2022 of $99.78 billion, is supported mainly by their increasingly use in the building and construction sec-tors, insulation and energy saving, weight reduction applications (e.g., automobile industry), as well as in the packaging end-use industry. This growing market can however face retraction related to the instability in the prices of raw materials. Nevertheless, the implemented R&D strategies at worldwide level, focused in the development of new raw materials and technologies, with focus on raw bio-based materials, are the main driving forces for the development of this area in the near future (MaMRP 2017). In this context, RPU foams, owned to their outstanding thermal-insulating properties, are considered materials with excellent performance. In fact, thermal insulation of RPU foams is known to be better than the one of other conventional materials (e.g., expanded polystyrene, mineral wool, cork, softwood, fireboard, concrete blocks, and bricks). These materials, although less expensive, require higher amounts to attain the same insulating performance. Other important characteristics of RPU foams are the high mechanical strength, adhesive properties, and easy processing (Lee and Ramesh 2004).

A typical RPU foam formulation includes an isocyanate, a polyol, a co-crosslinking agent, physical and/or chemical blowing agents, a catalyst, and a surfactant. The reactions involved in this process include urethane formation, crosslinking reactions, and foaming reactions. The more commonly used isocyanates are toluene diisocyanate (TDI) and polymeric MDI (PMDI). The used NCO index is usually comprised between 1.05 and 1.20, which corresponds to an NCO excess of 5–20% (Lee and Ramesh 2004; Sonnenschein 2015). The used polyols are primary and secondary hydroxyl-terminated polyether polyols followed by polyester polyols, including aromatic polyester polyols. These polyols have normally a high hydroxyl index, generally comprised between 300 and 800 mg KOH/g (Mihail 2005).

Most of the pioneering work dedicated to the incorporation of oxypropylated lignins into RPU foams has been performed by Gandini and co-workers. The produced RPU foams were found to have insulating properties, dimensional stability, and an accelerated aging behavior very similar to those prepared with commercial counterparts (Gandini et al. 2002; Nadji et al. 2005). In particular, in the work of Nadji et al. (2005), several foam formulations were produced by substituting a commercial polyol, partially or totally, by the lignin-based polyols (soda lignin, organosolv lignin, kraft lignin, and oxidized organosolv lignin were oxypropylated for this purpose). Foams prepared from oxypropylated organosolv lignin and oxypropylated soda lignin presented insulating properties, dimensional stability, and resistance against natural and accelerated aging, comparable to those prepared with the selected commercial polyol (reference foam). In contrast, oxypropylated kraft lignin and oxypropylated oxidized lignin give rise to foam formulations with non-suitable properties. More recently, Berrima et al. (2016) also used an oxypropylated soda lignin (precipitated from black liquor of *Stipa tenacissima L.* cooking) with interesting results. In brief, RPU foams with density ranging between 40 and 70 kg/m3, an average cell size of 0.4–0.8 mm, and compression modulus of 0.1–8.7 MPa were obtained, showing promising properties to be used in thermal and acoustic insulation applications.

Glasser and co-workers have also conducted studies dedicated to the synthesis of RPU foams from oxypropylated lignin but mainly concerned with the study of the flame-resistant properties of the produced cellular materials (Glasser and Leitheiser 1984). This topic was also treated in a more recent study, where oxyprolylated Biolignin™(organosolv lignin) was used to produce lignin-based RPU foams (Paberza et al. 2013) having in view the study of the flame-retardant properties and thermal stability. RPU foams were produced with the lignin-based polyols by replacing 0 to 100% of a commercial polyether polyol in the formulation. The flammability and thermal stability were studied and compared with a RPU foam prepared with lignin as a reactive filler (0–15%). The results have shown the improvement of flame resistance and thermal stability, both with the adding of the lignin-based polyol and lignin direct use. The optimal thermal and flammability properties were achieved where lignin-based polyols were used at a content of 10–15% (corresponding to the direct use of 8% lignin).

Lignin-based RPU foams using a sucrose-derived polyol and glycerol mixed with oxypropylated kraft pine lignin have been produced (Li and Ragauskas 2012b). The resulting foams presented low density (30 kg/m^3). Among the tested formulations, the one using solely the lignin-based polyol showed the best mechanical properties. In another study of the group (Li and Ragauskas 2012a), RPU foams with 100% of an ethanol organosolv lignin polyol were prepared and reinforced with cellulose nanowhiskers (CNWs) (1–5%, w/w). The produced lignin-based polyols were evaluated as suitable for RPU foams preparation. Moreover, the incorporation of CNWs significantly improved both mechanical and thermal properties. For both studies, the thermal conductivity of the prepared foams was not evaluated.

RPU foams using lignin-based polyols obtained by oxypropylation of four distinct technical lignins (Alcell, Indulin AT, Curan 27-11P, and Sarkanda) were produced in the work of Cateto et al. (2013). Polyol formulations using two L/PO/C ratios (g/ml/%w/w (biomass basis)) were chosen (30/70/2 and 20/80/5). RPU foams have been prepared with a polyol mixture comprising a commercial polyol (Lupranol® 3323) and the lignin-based polyol at contents ranging from 25% to 100% (w/w). Foams produced using the chosen polyol formulations exhibit properties similar to those of the reference foam (foam without adding the lignin-based polyol). Exceptions were registered, particularly with the foams produced with Sarkanda and Curan 27-11P-based polyols, which were found to present limitations for RPU foam formulations' purposes. RPU foams produced with Alcell and Indulin AT 30/70/2 polyols exhibited the best properties; also, they correspond to the samples incorporating the highest lignin content among the studied ones.

3.4   Lignin Use After Depolymerization

Regarding the highly random and branched three-dimensional polyphenolic structure of lignin, their partial disassembling would improve miscibility or compatibility with other ingredients. Lignin depolymerization results in the formation of a broad range of products, due to differences in lignin sources and used depolymerization methods (Abdelaziz et al. 2016; Xin et al. 2014; Zhu et al. 2017b). Generally, lignin depolymerization results in products with low molecular weight and high aliphatic/total hydroxyl content. This functionality improvement makes these lignin derivatives suitable comonomers to be used in the preparation of RPU foams providing a high degree of replacement for petrochemical-derived polyol. The major objective of all the fragmentation/depolymerization strategies is to cleave the ether linkages of the lignin structure. Chemical depolymerization of lignin can be divided into several categories according to the chemistry involved in the depolymerization process (Mahmood et al. 2016b): (i) acid-catalyzed, (ii) metallic catalyzed, (ii) ionic liquid-assisted, (iv) sub- or supercritical fluid-assisted, (v) base-catalyzed, and (vi) oxidative depolymerization. Each process presents some advantages and disadvantages.

After depolymerization, the access to the hydroxyl groups of the liquefied lignin can still be limited, constraint that can be overcome by chemical modification (e.g., oxypropylation) (Mahmood et al. 2016b). This strategy was followed by Mahmood and co-workers (2016a) who studied the liquefaction of a hydrolysis lignin (HL) using a 50:50 (v/v) water:ethanol mixture. The obtained product (powder form) was thereafter oxypropylated to produce liquid polyols with a bio-content of 50–70 wt%, which were used to prepare RPU foams. All the produced RPU foams were thermally stable up to 200 °C, exhibit high compression strengths, and presented thermal conductivity between 0.029 W/m.K and 0.034 W/m.K. In another work of the group (Mahmood et al. 2015), RPU foams were prepared following two different approaches: (i) by directly replacing PPG400 or sucrose polyol with depolymerized lignin and (ii) by using oxypropylated DKL as a single polyol feedstock. Among the studied routes, the one using oxypropylated depolymerized lignins gives rise to foams with a superior combination of physical, mechanical, and thermal properties. All the foams showed good compression strengths and thermal conductivity between 0.029 W/m K and 0.040 W/m K and were thermally stable up to 200 °C. In both studies, the produced foams using oxypropylated depolymerized lignin were evaluated as suitable candidates for insulation applications.

In a similar work, Fernandes and co-workers (2016b) studied the preparation of polyols through oxypropylation from a depolymerized kraft lignin and their use in RPU foam formulations. The depolymerized lignin was, in this case, obtained as a by-product of an oxidation process developed at the LA LSRE-LCM to produce vanillin and syringaldehyde. The obtained polyol was incorporated into a RPU foam formulation, alone or combined with 50% (w/w) of a commercial polyol using a NCO/OH molar ratio of 1.1. Even lignin is referred in literature as presenting fire retardancy, leading to the formation of char and volatile compounds from thermal decomposition, its incorporation on RPU foam systems resulted only in slight advantages; nevertheless the HRR decreased for the systems using the depolymerized lignin fractions. In fact, and comparatively with the original lignin, these fractions showed interesting characteristics for polymeric materials production. An additional advantage was the elimination of the malodorous typical of kraft lignins, possible resulting from the loss of the side chains of phenylpropane unit (bounded to sulfhydryl moieties, in the case of kraft lignin). The evaluated mechanical properties and thermal conductivity pointed out their compliance for insulation applications.

Xue and co-workers (2017) evaluated the depolymerization of corncob lignin using an isopropanol-water mixture and NaOH as catalyst, having in view the production of biopolyols with low molecular weight and with appropriated hydroxyl number for the synthesis of a bio-based rigid polyurethane foams. The results showed that 22.1% of the depolymerized lignin presented a low molecular weight (860 g/mol), a suitable hydroxyl number (5.10 mmol/g), and a solid residue of 5.1%. Based on the obtained results, the authors concluded that depolymerization process is appropriated for the production of biopolyols with suitable properties to partially or totally replace oil-based polyols in RPU foam formulations.

In a different approach, Arshanitsa and co-workers (2016) studied the incorporation of an organosolv wheat straw lignin (Biolignin™) into PU formulations. In order to increase lignin solubility, lignin was fractionated by sequential extraction with dichloromethane, methanol, and a mixture of both solvents. The total yield of the isolated soluble fractions was 40% (w/w). These individual fractions were characterized concerning the reactivity toward MDI. Moreover, PU films were produced by casting, from mixtures of the selected lignin fraction, PEG400 and MDI in tetrahydrofuran (THF) using a NCO/OH molar ratio of 1.05. The incorporated lignin content in the PU films varied from 5% to 40% (w/w). The obtained results pointed out that the isolated fractions of Biolignin™ acted as crosslinker agents in the PU network. The PU tensile properties were dependent of the type of lignin fraction and content, once the generated PU films were high elastic to glassy crosslinked materials. Even this approach differs from the ones based on depolymerization, it highlights the advantages of using lignin fractions with more controlled properties, instead of the whole original lignin. This approach, together with the use of depolymerized fractions, is currently a driving force toward the development of novel and more effective lignin-based materials.

References

Abdelaziz OY et al (2016) Biological valorization of low molecular weight lignin. Biotechnol Adv 34:1318–1346

Amaral JS, Sepúlveda M, Cateto CA, Fernandes IP, Rodrigues AE, Belgacem MN, Barreiro MF (2012) Fungal degradation of lignin-based rigid polyurethane foams. Polym Degrad Stab 97:2069–2076

Arbenz A, Avérous L (2015) Oxyalkylation of Gambier tannin - synthesis and characterization of ensuing biobased polyols. Ind Crop Prod 67:295–304

Arshanitsa A, Vevere L, Telysheva G, Dizhbite T, Gosselink RJA, Bikovens O, Jablonski A (2015) Functionality and physico-chemical characteristics of wheat straw lignin, biolignin™, derivatives formed in the oxypropylation process. Holzforsch. - Int J Biol Chem Phys Technol Wood 69:785–793

Arshanitsa A, Krumina L, Telysheva G, Dizhbite T (2016) Exploring the application potential of incompletely soluble organosolv lignin as a macromonomer for polyurethane synthesis. Ind Crop Prod 92:1–12

Avelino F, Almeida SL, Duarte EB, Sousa JR, Mazzetto SE, Souza Filho MSM (2017) Thermal and mechanical properties of coconut shell lignin-based polyurethanes synthesized by solvent-free polymerization. J Mater Sci 53:1470–1486

Balat M (2008) Mechanisms of thermochemical biomass conversion processes. Part 3: reactions of liquefaction. Energy Sources, Part A 30:649–659

Barbosa M, Matos M, Barreiro MF (2013) Gandini a chemical modifications as a strategy to modulate properties of Oxypropylated products. In: 4th green chemistry and nanotechnologies in polymer chemistry, Pisa, Italy, September 4–6

Belgacem NM, Gandini A (2008) Monomers, polymers and composites from renewable resources, 1st edn. Elsevier Science, Amsterdam

Bernardini J, Anguillesi I, Coltelli M-B, Cinelli P, Lazzeri A (2015) Optimizing the lignin based synthesis of flexible polyurethane foams employing reactive liquefying agents. Polym Int 64:1235–1244

Berrima B, Mortha G, Boufi S, Aloui EE, Mohamed BN (2016) Oxypropylation of soda lignin: characterization and application in polyurethanes foams production. Cellul Chem Technol 50:941–950

Borrero-López AM, Santiago-Medina FJ, Valencia C, Eugenio ME, Martin-Sampedro R, Franco JM (2017a) Valorization of Kraft lignin as thickener in Castor oil for lubricant applications. J Renew Mater 6:347

Borrero-López AM, Valencia C, Franco JM (2017b) Rheology of lignin-based chemical oleogels prepared using diisocyanate crosslinkers: effect of the diisocyanate and curing kinetics. Eur Polym J 89:311–323

Briones R, Serrano L, Younes RB, Mondragon I, Labidi J (2011) Polyol production by chemical modification of date seeds. Ind Crop Prod 34:1035–1040

Brown ME, Chang MC (2014) Exploring bacterial lignin degradation. Curr Opin Chem Biol 19:1–7

Cachet N, Camy S, Benjelloun-Mlayah B, Condoret J-S, Delmas M (2014) Esterification of organosolv lignin under supercritical conditions. Ind Crop Prod 58:287–297

Camarero S, Martínez MJ, Martínez AT (2014) Understanding lignin biodegradation for the improved utilization of plant biomass in modern biorefineries. Biofuels Bioprod Biorefin 8:615–625

Carriço CS, Fraga T, Pasa VMD (2016) Production and characterization of polyurethane foams from a simple mixture of castor oil, crude glycerol and untreated lignin as bio-based polyols. Eur Polym J 85:53–61

Cateto CA (2008) Lignin-based polyurethanes: characterisation, synthesis and applications. PhD thesis, Faculty of Engineering, Porto University

Cateto C, Barreiro MF, Rodrigues AE, Belgacem MN (2009) Optimization study of lignin oxypropylation in view of the preparation of polyurethane rigid foams. Ind Eng Chem Res 48:2583–2589

Cateto CA, Barreiro MF, Rodrigues AE, Belgacem MN (2011) Kinetic study of the formation of lignin-based polyurethanes in bulk. React Funct Polym 71:863–869

Cateto CA, Barreiro MF, Ottati C, Lopretti M, Rodrigues AE, Belgacem MN (2013) Lignin-based rigid polyurethane foams with improved biodegradation. J Cell Plast 50:81–95

Chauhan M, Gupta M, Singh B, Singh AK, Gupta VK (2014) Effect of functionalized lignin on the properties of lignin–isocyanate prepolymer blends and composites. Eur Polym J 52:32–43

Chen Y, Stark N, Cai Z (2014) Chemical modification of Kraft lignin: effect on chemical and thermal properties. Bioresources 9:5488–5500

Cheradame H, Detoisien M, Gandini A, Pla F, Roux G (1989) Polyurethane from kraft lignin. Br Polym J 21:269–275

Ciemniecki SL, Glasser WG (1988) Multiphase materials with lignin: 2. Blends of hydroxypropyl lignin with poly(vinyl alcohol). Polymer 29:1030–1036

Cui C, Sadeghifar H, Sen S (2013) Toward thermoplastic lignin polymers; part II: thermal & polymer characteristics of Kraft lignin & derivatives. Bioresources 8:864–886

Datta R, Kelkar A, Baraniya D, Molaei A, Moulick A, Meena R, Formanek P (2017) Enzymatic degradation of lignin in soil: a review. Sustainability 9:1163

D'Souza J, Camargo R, Yan N (2017) Biomass liquefaction and Alkoxylation: a review of structural characterization methods for bio-based polyols. Polym Rev 57:668–694

Duval A, Lawoko M (2014) A review on lignin-based polymeric, micro- and nano-structured materials. React Funct Polym 85:78–96

Evtiouguina M, Barros AM, Cruz-Pinto JJ, Neto CP, NaceurBelgacem CP, Gandini A (2000) The Oxypropylation of Cork residues: preliminary results. Bioresour Technology 73:187–189

Evtiouguina M, Barros-Timmons A, Cruz-Pinto JJ, Neto CP, Belgacem MN, Gandini A (2002) Oxypropylation of Cork and the use of the ensuing polyols in polyurethane formulations. Biomacromolecules 3:57–62

Evtuguin D, Andreolety J, Gandini A (1998) Polyurethanes based on oxygen-organosolv lignin. Eur Polym J 34:1163–1169

Fernandes IP, Barbosa M, Pinto V, Ferreira MJ (2014) Barreiro MF flexible polyurethane foams modified with lignin-based fillers for active footwear insoles. In: 5th workshop green chemistry and nanotechnologies in polymer chemistry, San Sebastian, Spain, July 9–11

Fernandes IP, Barbosa M, Amaral JS, Pinto V, Rodrigues JL, Ferreira MJ, Barreiro MF (2016a) Biobased additives as biodegradability enhancers with application in TPU-based footwear components. J Renew Mater 4:47–56

Fernandes IP et al (2016b) Polyols and rigid polyurethane foams derived from lignin side-streams of the pulp and paper industry. In: XXIII TECNICELPA international conference, Porto, Portugal, October 12–14

Gandini A (2011) The irruption of polymers from renewable resources on the scene of macromolecular science and technology. Green Chem 13:1061

Gandini A, Lacerda TM (2015) From monomers to polymers from renewable resources: recent advances. Prog Polym Sci 48:1–39

Gandini A, Belgacem MN, Guo Z-X, Montanari S (2002) Lignins as macromonomers for polyesters and polyurethanes. In: Hu TQ (ed) Chemical modification, properties, and usage of lignin. Springer US, Boston, pp 57–80

Glasser WG, Leitheiser RH (1984) Engineering plastics from lignin. Polym Bull 12:1–5

Glasser WG, Wu LCF, Selin JF (1983) Synthesis, structure, and some properties of Hydroxypropyl Lignins. In: Soltes EJ (ed) Wood a agricultural residues. Academic Press, New York, pp 149–166

Glasser WG, Barnett CA, Rials TG, Saraf VP (1984) Engineering plastics from lignin II. Characterization of hydroxyalkyl lignin derivatives. J Appl Polym Sci 29:1815–1830

Glasser WG, Oliveira WD, Stephen S. Kelley, Nieh LS (1990) Method of producing prepolymers from hydroxyalkyl lignin derivatives. USA Patent

Glasser W, Oliveira W, Kelley S, Niehl S (1991) Method of producing star-like polymers from Lignin. USA Patent

Gómez-Fernández S, Ugarte L, Calvo-Correas T, Peña-Rodríguez C, Corcuera MA, Eceiza A (2017) Properties of flexible polyurethane foams containing isocyanate functionalized Kraft lignin. Ind Crop Prod 100:51–64

Grandviewreport (2017) Lignin market analysis by product (Ligno-sulphonates, Kraft Lignin, Organosolv Lignin) By application (Macromolecules, Aromatics), by region (North America, Europe, APAC, Central & South America, MEA), and segment forecasts, 2014–2025

Hatakeyama H (2002) Polyurethanes containing lignin. In: Hu TQ (ed) Chemical modification, properties, and usage of lignin. Springer US, Boston, pp 41–56

Hatakeyama H, Kosugi R, Hatakeyama T (2008) Thermal properties of lignin and molasses-based polyurethane foams. J Therm Anal Calorim 92:419–424

Hatakeyama H, Hirogaki A, Matsumura H, Hatakeyama T (2013) Glass transition temperature of polyurethane foams derived from lignin by controlled reaction rate. J Therm Anal Calorim 114:1075–1082

Hirose S, Yano S, Hatakeyama T, Hatakeyama H (1989) Heat-resistant polyurethanes from Solvolysis lignin. In: Glasser WG, Sarkanen S (eds) Lignin, vol 397. ACS symposium series. American Chemical Society, Washington, DC, pp 382–389

Hofmann K, Glasser WG (1993) Engineering plastics from lignin. 21.1Synthesis and properties of Epoxidized lignin-poly (propylene oxide) copolymers. J Wood Chem Technol 13:73–95

Holladay JE, Bozell JJ, White JF, Johnson D (2007) Top value-added chemicals from biomass. Volume II - Results of screening for potential. Candidates from Biorefinery Lignin. PNNL-16983

Ignat L, Ignat M, Ciobanu C, Doroftei F, Popa VI (2011a) Effects of flax lignin addition on enzymatic oxidation of poly(ethylene adipate) urethanes. Ind Crop Prod 34:1017–1028

Ignat L, Ignat M, Stoica A, Ciobanu C, Popa V (2011b) Lignin blends with polyurethane-containinng lactate segments. Properties and enzymatic degradation effects. Cellul Chem Technol 45:233–243

Jasiukaitytė-Grojzdek E, Kunaver M, Crestini C (2012) Lignin structural changes during liquefaction in acidified ethylene glycol. J Wood Chem Technol 32:342–360

Jeong H, Park J, Kim S, Lee J, Ahn N (2013) Compressive viscoelastic properties of softwood Kraft lignin-based flexible polyurethane foams. Fibers Polym 14:1301–1310

Jin Y, Ruan X, Cheng X, Lu Q (2011) Liquefaction of lignin by polyethyleneglycol and glycerol. Bioresour Technol 102:3581–3583

Kai D, Tan MJ, Chee PL, Chua YK, Yap YL, Loh XJ (2016) Towards lignin-based functional materials in a sustainable world. Green Chem 18:1175–1200

Kumari S, Chauhan GS, Monga S, Kaushik A, Ahn J-H (2016) New lignin-based polyurethane foam for wastewater treatment. RSC Adv 6:77768–77776

Luo H, Abu-Omar MM (2017) Chemicals from lignin. In: Abraham, MA, Elsevier, Amsterdam, pp 573–585

Lange J-P (2007) Lignocellulose conversion: an introduction to chemistry, process and economics. Biofuels Bioprod Biorefin 1:39–48

Lee ST, Ramesh NS (2004) Polymeric foams - mechanisms and materials polymeric foams. CRC Press, Boca Raton

Li Y, Ragauskas AJ (2012a) Ethanol organosolv lignin-based rigid polyurethane foam reinforced with cellulose nanowhiskers. RSC Adv 2:3347

Li Y, Ragauskas AJ (2012b) Kraft lignin-based rigid polyurethane foam. J Wood Chem. Technol 32:210–224

Li S, Guo G, Nan X, Ma Y, Ren S, Han S (2013) Selective liquefaction of lignin from bio-ethanol production residue using Furfuryl alcohol. Bioresources 8:4563–4573

Li Y, Luo X, Hu S (2015) Bio-based polyols and polyurethanes. Green chemistry for sustainability. Springer International Publishing AG

Li H et al (2017a) High modulus, strength, and toughness polyurethane elastomer based on unmodified lignin. ACS Sustain Chem Eng 5:7942–7949

Li J, Wang B, Chen K (2017b) The use of lignin as cross-linker for polyurethane foam for potential application in adsorbing materials. Bioresources 12:853–867

Liew CY, Husaini A, Hussain H, Muid S, Liew KC, Roslan HA (2011) Lignin biodegradation and ligninolytic enzyme studies during biopulping of Acacia mangium wood chips by tropical white rot fungi. World J Microbiol Biotechnol 27:1457–1468

Long J, Lou W, Wang L, Yin B, Li X (2015) [$C_4H_8SO_3$Hmim]HSO_4 as an efficient catalyst for direct liquefaction of bagasse lignin: decomposition properties of the inner structural units. Chem Eng Sci 122:24–33

Lora JH, Glasser WG (2002) Recent industrial applications of lignin: a sustainable alternative to nonrenewable materials. J Polym Environ 10:39–48

Luo X, Mohanty A, Misra M (2013) Lignin as a reactive reinforcing filler for water-blown rigid biofoam composites from soy oil-based polyurethane. Ind Crop Prod 47:13–19

MaMRP (2017) Rigid foam market by type (Polyurethane, Polystyrene, Polyethylene, Polypropylene, Polyvinyl-Chloride), end-use industry (Building & Construction, Appliances, Packaging, Automotive), Region - Global Forecast to 2022

Mihail I (2005) Chemistry and technology of polyols for polyurethanes, 1st edn. Rapra Technology Limited, Shropshire, UK

Mahmood N, Yuan Z, Schmidt J, Xu C (2015) Preparation of bio-based rigid polyurethane foam using hydrolytically depolymerized Kraft lignin via direct replacement or oxypropylation. Eur Polym J 68:1–9

Mahmood N, Yuan Z, Schmidt J, Tymchyshyn M, Xu C (2016a) Hydrolytic liquefaction of hydrolysis lignin for the preparation of bio-based rigid polyurethane foam. Green Chem 18:2385–2398

Mahmood N, Yuan Z, Schmidt J, Xu C (2016b) Depolymerization of lignins and their applications for the preparation of polyols and rigid polyurethane foams: a review. Renew Sust Energ Rev 60:317–329

Matos M, Barreiro MF, Gandini A (2010) Olive stone as a renewable source of biopolyols. Ind Crop Prod 32:7–12

Matsushita Y (2015) Conversion of technical lignins to functional materials with retained polymeric properties. J Wood Sci 61:230–250

Montanari S, Baradie B, Andréolèty JP, Gandini A (1996) Star-shaped and crosslinked polyure-
thanes derived from lignins and oligoether isocyanates In: Kennedy JF, Phillips GO, Williams
PA (eds) The chemistry and processing of wood and plant fibrous material. Woodhead
Publishing, Cambridge, pp 351–358
Muller LC, Marx S, Vosloo HCM (2017) Polyol preparation by liquefaction of technical Lignins
in crude glycerol. J Renew Mater 5:67–80
Nadji H, Bruzzèse C, Belgacem MN, Benaboura A, Gandini A (2005) Oxypropylation of Lignins
and preparation of rigid polyurethane foams from the ensuing polyols. Macromol Mater Eng
290:1009–1016
Ni P, Thring R (2003) Synthesis of polyurethanes from solvolysis lignin using a polymerization
catalyst: mechanical and thermal properties. Int J Polym Mater 52:685–707
Norgren M, Edlund H (2014) Lignin: recent advances and emerging applications. Curr Opin
Colloid Interface Sci 19:409–416
Oliveira W, Glasser WG (1994) Multiphase materials with lignin. 11. Starlike copolymers with
caprolactone. Macromolecules 27:5–11
Oliveira W, Glasser WG (1994) Multiphase materials with lignin: 13. Block copolymers with cel-
lulose propionate. Polymer 35:1977–1985
Oliveira F, Ramires EC, Frollini E, Belgacem MN (2015) Lignopolyurethanic materials based on
oxypropylated sodium lignosulfonate and castor oil blends. Ind Crop Prod 72:77–86
Paberza A, Cabulis U, Arshanitsa A (2013) Flammability and thermal properties of rigid polyurethane foams containing wheat straw lignin. In: 4th workshop green chemistry and nanotechnologies in polymer chemistry, Pisa, Italy, September 4–6
Paberza A, Cabulis U, Arshanitsa A (2014) Wheat straw lignin as filler for rigid polyurethane
foams on the basis of tall oil amide. Polymery 59:34–38
Pan X, Saddler JN (2013) Effect of replacing polyol by organosolv and Kraft lignin on the property
and structure of rigid polyurethane foam. Biotechnol Biofuels 6:2–10
Panesar SS, Jacob S, Misra M, Mohanty AK (2013) Functionalization of lignin: fundamental stud-
ies on aqueous graft copolymerization with vinyl acetate. Ind Crop Prod 46:191–196
Pavier C, Gandini A (2000) Oxypropylation of sugar beet pulp. 1. Optimisation of the reaction.
Ind Crop Prod 12:1–8
Ragauskas AJ et al (2014) Lignin valorization: improving lignin processing in the biorefinery.
Science 344:1246843
Reimann A, Mörck R, Yoshida H, Hatakeyama H, Kringstad KP (1990) Kraft lignin in polyure-
thanes. III. Effects of the molecular weight of PEG on the properties of polyurethanes from a
Kraft lignin–PEG–MDI system. J Appl Polym Sci 41:39–50
Renata MS, Celeghini FM (2001) Optimization of the direct liquefaction of lignin obtained from
sugar cane bagasse. Energy Sources 23:369–375
Silva SHF, Santos PSB, Thomas da Silva D, Briones R, Gatto DA, Labidi J (2017) Kraft lignin-based
polyols by microwave: optimizing reaction conditions. J Wood Chem Technol 37:343–358
Sonnenschein MF (2015) Polyurethanes: science, technology, markets, and trends. Wiley series on
polymer engineering and technology. John Wiley & Sons, Inc., New Jersey
Sadeghifar H, Cui C, Argyropoulos DS (2012) Toward thermoplastic lignin polymers. Part
1. Selective masking of phenolic hydroxyl groups in Kraft Lignins via methylation and
Oxypropylation chemistries. Ind Eng Chem Res 51:16713–16720
Sanchez C (2009) Lignocellulosic residues: biodegradation and bioconversion by fungi. Biotechnol
Adv 27:185–194
Santos OS, Coelho da Silva M, Silva VR, Mussel WN, Yoshida MI (2017) Polyurethane foam
impregnated with lignin as a filler for the removal of crude oil from contaminated water.
J Hazard Mater 324:406–413
Saraf VP, Glasser WG, Wilkes GL (1985) Engineering plastics from lignin. VII. Structure property
relationships of poly(butadiene glycol)-containing polyurethane networks. J Appl Polym Sci
30:3809–3823
Sen S, Patil S, Argyropoulos DS (2015) Methylation of softwood Kraft lignin with dimethyl carbonate. Green Chem 17:1077–1087

Sequeiros A, Serrano L, Briones R, Labidi J (2013) Lignin liquefaction under microwave heating. J Appl Polym Sci 130:3292–3298

Serrano L, Alriols MG, Briones R, Mondragon I, Labidi J (2010) Oxypropylation of rapeseed cake residue generated in the biodiesel production process. Ind Eng Chem Res 49:1526–1529

Silva EAB, Zabkova M, Araújo JD, Cateto CA, Barreiro MF, Belgacem MN, Rodrigues AE (2009) An integrated process to produce vanillin and lignin-based polyurethanes from Kraft lignin. Chem Eng Res Des 87:1276–1292

Stecker F et al. (2014) Method for obtaining vanillin from aqueous basic compositions containing vanillin. WO Patent 2014006108 A1

Suman SK, Dhawaria M, Tripathi D, Raturi V, Adhikari DK, Kanaujia PK (2016) Investigation of lignin biodegradation by Trabulsiella sp. isolated from termite gut. Int Biodeter Biodegr 112:12–17

Tavares LB, Boas CV, Schleder GR, Nacas AM, Rosa DS, Santos DJ (2016) Bio-based polyurethane prepared from Kraft lignin and modified castor oil. Express Polym Lett 10:927–940

Ten E, Vermerris W (2015) Recent developments in polymers derived from industrial lignin. J Appl Polym Sci 132:42069–42081

Thring R, Vanderlaan M, Griffin S (1997) Polyurethanes from Alcell® lignin. Biomass Bioenergy 13:125–132

Thring RW, Ni P, Aharoni SM (2004) Molecular weight effects of the soft segment on the ultimate properties of lignin-derived polyurethanes. Int J Polym Mater 53:507–524

Upton BM, Kasko AM (2016) Strategies for the conversion of lignin to high-value polymeric materials: review and perspective. Chem Rev 116:2275–2306

Vanderlaan M, Thring RW (1998) Polyurethane from Alcell lignin fractions obtained by sequential solvent extraction. Biomass Bioenergy 14:525–531

Wu LCF, Glasser WG (1984) Engineering plastics from lignin. I. Synthesis of hydroxypropyl lignin. J Appl Polym Sci 29:1111–1123

Xiao B, Suna X, Sun R (2001) The chemical modification of lignins with succinic anhydride in aqueous systems. Polym Degrad Stab 71:223–231

Xin J, Zhang P, Wolcott MP, Zhang X, Zhang J (2014) Partial depolymerization of enzymolysis lignin via mild hydrogenolysis over Raney nickel. Bioresour Technol 155:422–426

Xing W, Yuan H, zhang P, Yang H, Song L, Hu Y (2013) Functionalized lignin for halogen-free flame retardant rigid polyurethane foam: preparation, thermal stability, fire performance and mechanical properties. J Polym Res 20:234–245

Xue BL, Wen JL, Sun RC (2015) Producing lignin-based polyols through microwave-assisted liquefaction for rigid polyurethane foam production. Materials (Basel) 8:586–599

Xue B-L, Huang P-L, Sun Y-C, Li X-P, Sun R-C (2017) Hydrolytic depolymerization of corncob lignin in the view of a bio-based rigid polyurethane foam synthesis. RSC Adv 7:6123–6130

Yadav M, Yadav HS (2015) Applications of ligninolytic enzymes to pollutants, wastewater, dyes, soil, coal, paper and polymers. Environ Chem Lett 13:309–318

Yoshida H, Mörck R, Kringstad KP, Hatakeyama H (1987) Kraft lignin in polyurethanes I. Mechanical properties of polyurethanes from a Kraft lignin–polyether triol–polymeric MDI system. J Appl Polym Sci 34:1187–1198

Yoshida H, Mörck R, Kringstad KP, Hatakeyama H (1990) Kraft lignin in polyurethanes. II. Effects of the molecular weight of Kraft lignin on the properties of polyurethanes from a Kraft lignin–polyether triol–polymeric MDI system. J Appl Polym Sci 40:1819–1832

Zhang C, Wu H, Kessler MR (2015) High bio-content polyurethane composites with urethane modified lignin as filler. Polymer 69:52–57

Zhu H et al (2014) Preparation and characterization of flame retardant polyurethane foams containing phosphorus–nitrogen-functionalized lignin. RSC Adv 4:55271–55279

Zhu D, Zhang P, Xie C, Zhang W, Sun J, Qian WJ, Yang B (2017a) Biodegradation of alkaline lignin by bacillus ligniniphilus L1. Biotechnol Biofuels 10:44

Zhu G, Jin D, Zhao L, Ouyang X, Chen C, Qiu X (2017b) Microwave-assisted selective cleavage of C α C β bond for lignin depolymerization. Fuel Process Technol 161:155–161

Chapter 4
Polyphenols from Bark of *Eucalyptus globulus*

Abstract *E. globulus* bark is extremely rich in polyphenolic compounds (polar fraction) being mainly composed of simple phenolic compounds, ellagitannins and proanthocyanidins. Several of these compounds exhibit important biological activities such as antioxidant, antimicrobial, anti-inflammatory activities. Moreover, they have interesting properties as ingredients for cosmetic formulations and as plasticizer agents and can be employed in leather tanning and adhesive and coating production, among other applications.

Pulp and paper industries generate large amounts of bark as a byproduct currently burned at the mill to produce power. However, aiming the valorization of pulp and paper side streams and integration of the biorefinery concept and circular economy, the extraction of polyphenolic compounds from *E. globulus* bark is a feasible option to increase the portfolio of products.

Several studies regarding the extraction of polyphenols from *E. globulus* bark employed water, methanol/ethanol, or aqueous alkaline mixtures as the extracting medium. The extraction of this family of compounds employing aqueous ethanol mixtures has been optimized by experimental design and two interesting extraction conditions established: a first one for the extraction optimization of polyphenolic compounds, quantified by Folin-Ciocalteu method (264 min, ethanol/water 52:48 v/v, 82.5 °C) and a second one showing the best activity against breast cancer cells (360 min, ethanol/water 80:20 v/v, 82.5 °C). Membrane separation with polymeric membranes and adsorption onto a nonpolar adsorbent studies performed with the ethanolic extracts selected have demonstrated that these technologies can be applied to further concentrate the polyphenolic fraction of the extracts and improve its biological properties.

Keyword Composition of polar extracts · Polyphenols · Water and alkaline extraction · Ethanol/water extraction · Fractionation · Membrane processing · Diafiltration and adsorption

© Springer Nature Switzerland AG 2018

A. E. Rodrigues et al., *An Integrated Approach for Added-Value Products from Lignocellulosic Biorefineries*, https://doi.org/10.1007/978-3-319-99313-3_4

4.1 Composition of Polar Extracts

The chemical composition of *E. globulus* bark was disclosed in Sect. 1.6, and it was referred that the bark extractives are mainly of polar nature accounting for 80–86% of the total extractives (Miranda et al. 2012, 2013; Neiva et al. 2014). Several studies have been conducted and demonstrated that polar extracts of *E. globulus* bark can be an important source for phenolic compounds. The presence of numerous phenolic compounds in *Eucalyptus* barks has been reported in several studies (Cadahía et al. 1997; Conde et al. 1995, 1996; Fechtal and Riedl 1991; Santos et al. 2011; Yazaki and Hillis 1976), and herein the phenolic compounds identified in particular for *E. globulus* bark will be detailed.

The polyphenols or phenolic compounds are a complex group of aromatic plant metabolites that are widely distributed throughout the plant kingdom and possess important physiological and morphological functions in plants (Bravo 1998). In the specific case of barks, this group of compounds plays an important role in defense mechanisms (Balasundram et al. 2006).

The phenolic compounds are, in general, considered to be weak acids and possess at least one aromatic ring containing one or more hydroxyl groups. They can be categorized into several classes ranging from simple phenolic compounds (e.g., phenolic acids and aldehydes or flavonoids) to highly polymerized compounds (e.g., hydrolysable and condensed tannins or proanthocyanidins) (Bravo 1998). Sometimes, these compounds may occur as functional derivatives such as esters and methyl esters or can be present as conjugates with mono- and polysaccharides and are designated as glycosides.

The polyphenols are known for being bioactive compounds exhibiting antioxidant, anti-inflammatory, antibacterial, antimicrobial, antithrombotic, and anticarcinogenic activities (Bravo 1998; Harborne and Williams 2000; Hertog et al. 1992; Meltzer and Malterud 1997; Mukhatar et al. 1988; Vázquez et al. 2008). The antioxidant activity of the polyphenols is related with its capacity to scavenge free radicals, donate hydrogen atoms or electrons, or chelate metal cations (Balasundram et al. 2006; Moure et al. 2001; Rice-Evans et al. 1997; Stevanovic et al. 2009; Takuo 2005). Moreover, these compounds possess other appealing properties as natural specialty ingredients for food and beverage industries, nutrition, health, and personal cares industries (Pizzi 2008; Royer et al. 2013; Stevanovic et al. 2009; Zillich et al. 2015).

The polarity of the solvent used is detrimental for the type and amount of compounds extracted. The less polar compounds from the bark, including the less polar flavonoids such as isoflavones, flavones, methylated flavones, and flavonols, can be extracted with chloroform, dichloromethane, diethyl ether, or ethyl acetate, while the flavonoid glycosides and more polar aglycones are extracted with alcohols or alcohol-water mixtures. Catechins, proanthocyanidins, and condensed tannins can be extracted either with water, methanol, ethanol, or acetone (Marston and Hostettmann 2006).

Polar extracts of *E. globulus* bark are mainly composed of polyphenolic compounds, and several studies have identified the presence of phenolic acids and esters, flavonoids, flavonoid glycosides, gallolylglucose derivatives, and ellagitannins (Cadahía et al. 1997; Conde et al. 1995, 1996; Eyles et al. 2004; Kim et al. 2001; Santos et al. 2011; Vázquez et al. 2012; Yazaki and Hillis 1976). In Table 4.1 it is summarized the phenolic compounds that have already been identified in *E. globulus* bark, being several of them very valuable in the market and known for its important biological activities, as, for example, taxifolin, ellagic acid, apigenin, and naringenin.

The composition of the polar extracts of *E. globulus* bark has been studied by several authors, and most of the published literature is related with the identification of the polyphenolic fraction after bark extraction with methanol, water, or aqueous methanol (Cadahía et al. 1997; Conde et al. 1995, 1996; Santos et al. 2011; Yazaki and Hillis 1976). It is important to note that the studies dedicated with studying the composition of polar extracts of *Eucalyptus* usually encompass a previous extraction step with hexane or dichloromethane for the removal of the lipophilic fraction and the less polar compounds. This is the case of the study of Santos et al. (2011), where a previous extraction step of the *E. globulus* bark with dichloromethane is performed prior to the bark extraction with methanol or aqueous methanolic solution. On the contrary, other studies dedicated with the identification of the polar extractives of bark perform subsequent extraction(s) or partition(s) step(s) with *n*-hexane, ethyl ether, chloroform, n-butanol, and/or water (Cadahía et al. 1997; Conde et al. 1995, 1996; Kim et al. 2001; Yun et al. 2000) in order to concentrate and isolate the compounds.

The first study referring to the identification of polyphenols from polar extracts of *E. globulus* bark was performed by Yazaki and Hillis (1976) where several phenolic acids, including free ellagic and gallic acids, catechin, and methyl and glycosyl derivatives of ellagic acid, have been identified, including two ellagic acid derivatives, 3-methyl-ellagic acid glucoside and 3-*O*-methyl-ellagic acid 4′-rhamnoside, in small and large amounts, respectively.

Later, the presence of phenolic acids (e.g., gallic, protocatechuic, ellagic, and vanillic acids), protocatechuic aldehyde, and several flavonoids (e.g., taxifolin, eriodictyol, quercetin, and naringenin) was reported in methanol/water 80:20%v/v extracts of *E. globulus* bark (Cadahía et al. 1997; Conde et al. 1995, 1996). In Conde et al. (1996), high variability of extraction yields was obtained according to geographical location and environmental conditions. After the extraction of the polar compounds with methanol/water 80:20%v/v, the ether-soluble fraction was analyzed in detail regarding its composition in phenolic compounds. About 7–41% of the extracted compounds in methanol/water were soluble in ether revealing the presence of some polymeric phenolic compounds such as condensed tannins that were extracted with the methanolic solution and did not solubilized in ether. The extracts were mainly composed of gallic and ellagic acids, eriodictyol, and several ellagitannins. Other compounds were also detected such as protocatechuic acid,

Table 4.1 Summary of phenolic compounds identified in *E. globulus* bark extractives and abundance

Name of compound	Extraction medium	Amount mg g^{-1} extract	mg kg^{-1} dried bark	References
Phenolic acids, aldehydes, and esters				
Protocatechuic aldehyde	Methanol/water 80:20%v/v	Detected		Conde et al. (1996)
Protocatechuic acid	Water	2.8	54.0	Santos et al. (2011)
	Methanol	1.6	133.1	Santos et al. (2011)
	Methanol/water 80:20%v/v	Detected		Conde et al. (1996)
	Methanol/water 50:50%v/v	2.1	194.1	Santos et al. (2011)
Vanillic acid	Methanol/water 80:20%v/v	Trace amounts		Conde et al. (1996)
Quinic acid	Water	2.5	47.6	Santos et al. (2011)
	Methanol	1.5	120.3	Santos et al. (2011)
	Methanol/water 50:50%v/v	1.5	139.5	Santos et al. (2011)
Dihydroxyphenyl acetic acid	Water	0.4	7.3	Santos et al. (2011)
Chlorogenic acid	Water	5.2	101.0	Santos et al. (2011)
	Methanol	Detected		Yazaki and Hillis (1976)
	Methanol	6.0	492.9	Santos et al. (2011)
	Methanol/water 50:50%v/v	13.4	1239.5	Santos et al. (2011)
Chlorogenic acid isomer	Methanol	Detected		Yazaki and Hillis (1976)
Caffeic acid	Water	5.0	97.0	Santos et al. (2011)
Ellagic acid	Water	Trace amounts		Santos et al. (2011)
		Detected		Vázquez et al. (2012)
	Methanol	Detected		Yazaki and Hillis (1976)
		4.9	408.0	Santos et al. (2011)
	Methanol/water 80:20%v/v	Detected		Conde et al. (1996)
	Methanol/water 50:50%v/v	5.1	471.0	Santos et al. (2011)
Gallic acid	Water	3.5	67.9	Santos et al. (2011)
	Methanol	3.4	280.8	Santos et al. (2011)
		Detected		Yazaki and Hillis (1976)
	Methanol/water 80:20%v/v	Detected		Conde et al. (1996)
		Detected		Cadahía et al. (1997)
	Methanol/water 50:50%v/v	8.8	819.5	Santos et al. (2011)

(continued)

Table 4.1 (continued)

Name of compound	Extraction medium	Amount mg g^{-1} extract	mg kg^{-1} dried bark	References
Flavonoids				
Catechin	Water	15.9	307.8	Santos et al. (2011)
		Detected		Vázquez et al. (2012)
	Methanol	6.6	541.6	Santos et al. (2011)
		Detected		Yazaki and Hillis (1976)
	Methanol/water 80:20%v/v	Detected		Yun et al. (2000)
	Methanol/water 50:50%v/v	14.2	1320.6	Santos et al. (2011)
Epicatechin	Water	Detected		Vázquez et al. (2012)
Taxifolin	MeOH	1.5	121.7	Santos et al. (2011)
	Methanol/water 80:20%v/v	Detected		Conde et al. (1996)
		Detected		Yun et al. (2000)
	Methanol/water 50:50%v/v	7.8	721.6	Santos et al. (2011)
Mearnsetin	Methanol	0.3	27.8	Santos et al. (2011)
	Methanol/water 50:50%v/v	0.4	35.3	Santos et al. (2011)
Eriodictyol	Water	Trace amounts		Santos et al. (2011)
	Methanol	6.9	568.5	Santos et al. (2011)
	Methanol/water 80:20%v/v	Detected		Conde et al. (1996)
		Detected		Yun et al. (2000)
	Methanol/water 50:50%v/v	7.91	733.86	Santos et al. (2011)
Luteolin	Methanol	2.31	190.1	Santos et al. (2011)
	Methanol/water 50:50%v/v	3.66	340.0	Santos et al. (2011)
Quercetin	Methanol/water 80:20%v/v	Detected		Conde et al. (1995)
		Detected		Conde et al. (1996)
Isorhamnetin	Water	Detected		Vázquez et al. (2012)
	Methanol	3.98	327.8	Santos et al. (2011)
	Methanol/water 50:50%v/v	4.65	431.6	Santos et al. (2011)
Naringenin	Methanol	0.79	65.2	Santos et al. (2011)
	Methanol/water 80:20%v/v	Detected		Conde et al. (1995)
		Detected		Conde et al. (1996)
	Methanol/water 50:50%v/v	0.76	70.9	Santos et al. (2011)
Apigenin	Methanol/water 80:20%v/v	Detected		Conde et al. (1996)
Several polymeric proanthocyanidins	Methanol/water 80:20%v/v	Detected		Cadahía et al. (1997)

(continued)

Table 4.1 (continued)

Name of compound	Extraction medium	Amount mg g^{-1} extract	mg kg^{-1} dried bark	References
Flavonol	Methanol/water 80:20%v/v	Detected		Conde et al. (1995, 1996)
Rhamnazin	Methanol/water 80:20%v/v	Detected		Yun et al. (2000)
Rhamnetin	Methanol/water 80:20%v/v	Detected		Yun et al. (2000)
Flavonoid glycosides				
Isorhamnetin-hexoside	Water	Trace amounts		Santos et al. (2011)
	Methanol	1.53	126.4	Santos et al. (2011)
	Methanol/water 50:50%v/v	1.08	99.9	Santos et al. (2011)
Quercetin-hexoside	Methanol	0.15	12.2	Santos et al. (2011)
	Methanol/water 50:50%v/v	0.63	58.8	Santos et al. (2011)
Quercetin-3-*O*-rhanonoside	Water	Detected		Vázquez et al. (2012)
Myricetin-rhamnoside	Methanol	Trace amounts		Santos et al. (2011)
	Methanol/water 50:50%v/v	Trace amounts		Santos et al. (2011)
Isorhamnetin-rhamnoside	Water	0.17	3.29	Santos et al. (2011)
	Methanol	9.79	806.57	Santos et al. (2011)
	Methanol/water 50:50%v/v	10.00	927.64	Santos et al. (2011)
Aromadendrin-rhamnoside	Methanol	Trace amounts		Santos et al. (2011)
	Methanol/water 50:50%v/v	0.79	73.52	Santos et al. (2011)
Phloridzin	Methanol	Trace amounts		Santos et al. (2011)
	Methanol/water 50:50%v/v	0.75	69.87	Santos et al. (2011)
Mearnsetin-hexoside	Methanol	1.07	88.10	Santos et al. (2011)
	Methanol/water 50:50%v/v	1.30	121.12	Santos et al. (2011)
Engelitin	Methanol/water 80:20%v/v	Detected		Yun et al. (2000)
Hydrolysable tannins: gallotannins, ellagitannins, and derivatives				
Methyl gallate	Water	2.43	46.95	Santos et al. (2011)
	Methanol	0.68	56.18	Santos et al. (2011)
	Methanol/water 80:50%v/v	Detected		Yun et al. (2000)
	Methanol/water 50:50%v/v	1.50	139.25	Santos et al. (2011)
Bis(hexahydroxydiphenolyl)-glucose	Water	0.83	16.1	Santos et al. (2011)
	Methanol	0.68	56.4	Santos et al. (2011)
	Methanol/water 50:50%v/v	1.02	94.3	Santos et al. (2011)

(continued)

Table 4.1 (continued)

Name of compound	Extraction medium	Amount mg g^{-1} extract	mg kg^{-1} dried bark	References
Galloyl-bis(hexahydroxydiphenoyl)-glucose	Water	4.99	96.4	Santos et al. (2011)
	Methanol	7.23	595.9	Santos et al. (2011)
	Methanol/water 50:50%v/v	4.85	450.37	Santos et al. (2011)
Galloyl-hexahydroxydiphenoyl-glucose	Water	9.04	174.42	Santos et al. (2011)
	Methanol	9.27	764.15	Santos et al. (2011)
	Methanol/water 50:50%v/v	6.86	637.42	Santos et al. (2011)
Monogalloylglucose	Water	Detected		Vázquez et al. (2012)
Digalloylglucose	Water	6.35	122.64	Santos et al. (2011)
		Detected		Vázquez et al. (2012)
	Methanol	17.95	1479.4	Santos et al. (2011)
	Methanol/water 50:50%v/v	17.77	1648.9	Santos et al. (2011)
Tri-galloyl-glucose	Water	Detected		Vázquez et al. (2012)
Tetra- to tetradeca-galloyl-glucoses	Water	Detected		Vázquez et al. (2012)
3,4,5-Trimethoxyphenol −1-O-β-6´-O-galloyl glucopyranoside	Methanol/water 80:20%v/v	Detected		Yun et al. (2000)
Gallocatechin	Methanol	Detected		Yazaki and Hillis (1976)
Several ellagitannins	Methanol	Detected		Yazaki and Hillis (1976)
	Methanol/water 80:20%v/v	Detected		Conde et al. (1995)
		Detected		Conde et al. (1996)
		Detected		Cadahía et al. (1997)
Methyl-ellagic acid-pentose	Methanol	Trace amounts		Santos et al. (2011)
	Methanol/water 50:50%v/v	Trace amounts		Santos et al. (2011)
3-Methylellagic acid glucoside	Methanol	Detected		Yazaki and Hillis (1976)
3-O-Methylellagic acid 4´-rhamnoside	Methanol	Detected		Yazaki and Hillis (1976)
3-O-Methylellagic acid 3'-O-α-rhamnopyranoside	Methanol/water 80:20%v/v	Detected		Kim et al. (2001)
Several ellagic-acid-acetylrhamnosides	Methanol/water 80:20%v/v	Detected		Kim et al. (2001)
3-Methylellagic acid glucoside	Methanol	Detected		Yazaki and Hillis (1976)

protocatechuic aldehyde, quercetin, naringenin, apigenin, and taxifolin. Cadahía et al. (1997) have shown that the *E. globulus* bark is abundant in proanthocyanidins, and ellagitannins are abundant in *E. globulus* bark.

Kim et al. (2001) extracted the polar compounds of *E. globulus* stem bark with methanol/water 80:20%v/v solution and managed to isolate and identify four phenolic compounds with sequential partition with chloroform and ethyl acetate. The authors identified one compound as an ellagic acid rhamnoside and three ellagic acid acetylrhamnosides (3-*O*-methylellagic acid 3'-*O*-α-3"-*O*-acetylrhamonopyranoside, 3-*O*-methylellagic acid 3'-*O*-α-2"-*O*-acetylrhamono-pyranoside, 3-*O*-methylellagic acid 3'-*O*-α-4"-*O*-acetylrhamonopyranoside). The four compounds demonstrated reasonable antioxidant activities against lipid per-oxidation in rat liver microsomes.

In a similar study, Yun et al. (2000) managed to extract and isolate from *E. globu-lus* bark several phenolic compounds exhibiting significant to moderate antioxidant activity against peroxidation in rat liver microsomes. Rhamnazin, rhamnetin, erio-dictyol, quercetin, and taxifolin have significantly inhibited the lipid peroxidation induced by nonenzymatic Fe(II)-ascorbic acid system.

Recently, Santos et al. (2011) reported, among other polyphenols, 16 new pheno-lic compounds in *E. globulus* bark methanolic extracts, as, for example, quinic, dihydroxyphenylacetic, and caffeic acids, bis(hexahydroxydiphenoyl (HHDP))-glucose, galloyl-bis (HHDP), galloyl-(HHDP)-glucose, isorhamnetin-hexoside, mearnsetin, phloridzin, mearnsetin-hexoside, luteolin, and a proanthocyanidin B-type dimer. In this study polyphenols were identified and quantified after a sequential bark extraction with methanol and water, and with a methanol/water (50:50 v/v) solution. Similar extraction yields were obtained for both methanol and aqueous/methanol extracts of 9.28% and 8.24%, respectively. On the contrary, the extract obtained with water, after a previous extraction with methanol, exhibited significantly less extraction yield (1.93%), probably because the majority of the compounds were previously extracted with methanol. The same trend was observed regarding the polyphenols quantified by HPLC-UV where about 10851.52, 7279.34, and 1142.26 mg kg^{-1}_{bark} of polyphenols were quantified for methanol/water mixture, methanol, and water extracts, respectively. In terms of mg $g^{-1}_{extract}$ (i.e., selectivity of extraction), the content in total phenolic compounds identified was of 116.93, 88.34, and 59.18 mg $g^{-1}_{extract}$, for methanol/water mixture, methanol, and water extracts, respectively. In accordance, total phenolic compounds quantified by the Folin-Ciocalteu method were higher for methanol/water extract (413.8 mg_{GAE} $g^{-1}_{extract}$), immediately followed by methanol extract (409.7 mg_{GAE} $g^{-1}_{extract}$) and sig-nificantly less for the water extract (115.3 mg_{GAE} $g^{-1}_{extract}$).

In Table 4.1 it is summarized the amount of each phenolic compound quantified by HPLC-UV. In the aqueous/methanol extract, the main compounds identified were digalloylglucose, isorhamnetin-rhamnoside, and galloyl-HHDP-glucose, cor-responding to about 39% of the identified compounds. In the methanolic extract, about 42% of the identified compounds corresponded to digalloylglucose, catechin, and chlorogenic acid. The major components identified in the water extract were catechin, galloylhexahydroxyphenoyl-glucose, and digalloylglucose, representing about 53% of the identified compounds.

The polyphenol composition of *E. globulus* bark extracts obtained with supercritical fluid extraction employing CO_2 modified with 15% ethanol has also been studied (Santos et al. 2012). Detailed characterization of this extract was performed, and although the extraction yield and phenolic compounds' content estimated by the Folin-Ciocalteu method (0.32% and 33.10 mg_{GAE} $g^{-1}_{extract}$) was significantly lower than employing more polar solvents (e.g., 50% methanol/ethanol/water solutions, 9.28/9.79% and 407.41/159.58 mg_{GAE} $g^{-1}_{extract}$), the authors have observed that the CO_2 modified with ethanol was the most selective one for the extraction of eriodictyol, naringenin, and isorhamnetin, accounting for about 72% of all compounds identified by HPLC. In this work, methyl ellagic acid was identified for the first time.

Several polygalloyl-glucoses (gallotannins), catechin, epicatechin, ellagic acid, quercetin-3-*O*-rhamnoside, and isorhamnetin were identified in aqueous extracts of *E. globulus* bark by Vazquez et al. (2012), and these extracts revealed ferric reducing antioxidant power (FRAP) antioxidant activity.

4.2 Extraction of Polyphenols

Tree barks can be explored as raw material to extract several polyphenols with important applications. The extraction of polar compounds from *E. globulus* bark and characterization of the extracts obtained are addressed herein, mainly focused on applying water, alkaline solutions, ethanol, and its mixtures as the extracting media. Literature is extensive, and due to the natural heterogeneity of the plant material, data must be carefully compared. Moreover, the operating conditions for the extraction of polyphenolics, such as temperature, time of extraction, solvent used, and liquid-solid ratio, among others, also influence the quality and the yield of extraction obtained.

4.2.1 Water and Alkaline Extractions: Selectivity and Concentration Strategy

Table 4.2 summarizes the existing literature regarding the *E. globulus* bark extractions with water and alkaline solutions displaying the respective extraction yields, phenolic compound content (TPC), and formaldehyde-condensable tannins (fcT) obtained for different extraction conditions such as solid/liquid ratio, temperature, and time of extraction. TPC values are either expressed as %w/w$_{bark}$, which corresponds to a measure of the extraction yield of polyphenols, or as %w/w$_{extract}$, which is a measure of the extraction selectivity or extraction purity for polyphenols.

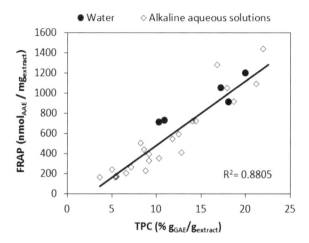

Fig. 4.1 FRAP content versus TPC in the extract

The fcT measures the formaldehyde-condensable polyphenol content assessing the potential of the extract for wood adhesives.

Several studies have demonstrated that TPC and FRAP antioxidant activity in extracts of different plants are linearly correlated (Maksimovic et al. 2005; Thaipong et al. 2006; Vázquez et al. 2009, 2012; Velioglu et al. 1998) probably explained by similarities of the methodologies employed (Folin-Ciacalteu and FRAP methods). For *E. globulus* bark extracts, positive correlations between both parameters are observed as well, with coefficients of determination between 0.658 and 0.8915 (Vázquez et al. 2008, 2012). The same linear behavior can be observed for the experiments summarized in Table 4.2, regardless of the solvent employed (Fig. 4.1), with a coefficient of determination 0.8805.

Regarding the use of water as the extracting medium, it is observed an almost linear trend between the bark total extractable content and the temperature increase, regardless of the other extraction conditions. This is somewhat expected since the temperature increase favors the diffusion of the solvent, the swelling of the matrix, and the solubility of compounds.

The highest total extraction yield is obtained at 140 °C of 17.7%w/w$_{dried\ bark}$. This trend is also observed for the extraction yield of TPC, with the exception of the experiments performed by Mota (2011), where the TPC yields obtained at room and boiling temperatures are somewhat higher than the values expected by following the linear trend. This deviation could be attributed to bark heterogeneity and to the different extracting times and solid/liquid ratios employed, which can influence the solubility of the phenolic compounds in solution and affect the extraction yield and/or degradation of the phenolic compounds since they are known for being thermal-sensitive.

In Mota et al. (2012) experimental design study, it was also obtained a linear response between the total extraction (Fig. 4.2a) and TPC extraction (Fig. 4.2b) yields as a function of time and temperature. It is possible to observe from the response

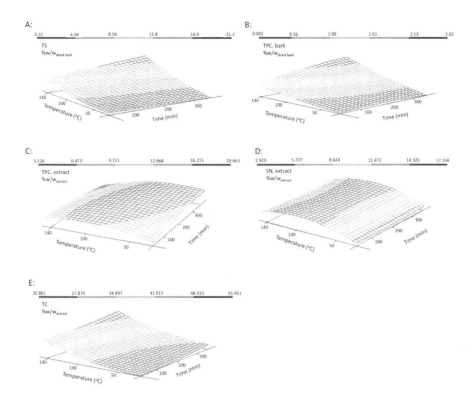

Fig. 4.2 Response surfaces obtained for aqueous *E. globulus* bark extractions employing water as extracting solvent. (**a**) Total solids, (**b**) extraction yield (%w/w$_{dried\ bark}$) of total phenolic compounds (TPC) and selectivity of extraction (%w/w$_{extract}$) of (**c**) TPC, (**d**) formaldehyde-condensable tannins given by Stiasny number (SN), and (E) total carbohydrates (TC)

surfaces that higher temperature and extraction times promote the extraction of compounds from bark and it is in accordance to the abovementioned observation.

Overall, considering all literature data, *E. globulus* bark water extractions performed at room temperature or 25 °C exhibit lower total extraction yields ranging from 1.93 to 2.45%w/w$_{dried\ bark}$ than the experiments performed at higher temperatures comprehended between 4.4 and 17.7%%w/w$_{dried\ bark}$. The same trend is observed for the extraction yields of the phenolic compounds (%g$_{GAE}$/g$_{dried\ bark}$) where yields of TPC have increased from 0.22% to 0.75% (at room T/25 °C) to 0.88–2.18% (100–140 °C).

In terms of selectivity of extraction of TPC with antioxidant activity in aqueous media, a negative effect of the temperature occurs because the phenolic compounds are thermal-sensitive and higher temperatures promote the co-extraction of other compounds. In the response surfaces obtained by Mota et al. (2012) regarding the selective extraction of TPC (Fig. 4.2c) and the particular family of polyphenols, fcT (Fig. 4.2d), it was well perceived that higher temperatures were prejudicial. Moreover, it was also observed that there is a trade-off between temperature and extracting time in order to selectively extract TPC. According to Fig. 4.2c, if lower

Proanthocyanidins (Pac) are particular class of polyphenols known for being bio-active compounds. In Table 4.3 it is summarized the literature data on Pac content obtained for some aqueous bark extractions where it is possible to observe that the extraction yield of these compounds is considerably lower regarding the overall extraction, therefore, allowing to conclude that water is not a good media to extract this type of polyphenols probably attributed to the hydrophobic character of this family of compounds. In fact, Mota et al. (2012) have demonstrated that the extraction of PAC is enhanced with ethanolic solutions and that water is a good medium to extract more hydrophilic compounds (e.g., carbohydrates).

Extractions with alkaline solutions have been applied focused on obtaining anti-oxidant compounds from several different matrixes (Vázquez et al. 2008, 2009, 2012) including *E. globulus* bark (Table 4.2).

In general, the extraction yields obtained employing alkaline solutions increase with the temperature increase, similarly to what was observed for experiments using water as the extracting medium. This statement has also been noted in Vazquez et al. (2009) study. Moreover, analyzing the total and TPC extraction yields obtained using alkali solutions, there is a greater dispersion of results attributed not only to the operational conditions such as extraction temperature, time, and solid/liquid ratio but also related to the type of salt employed and its concentration. Literature data contemplates the use of salts of NaOH, Na_2SO_3, and/or Na_2CO_3.

When alkaline solutions are employed as the extracting media instead of water, it is possible to retrieve from the literature data (Fig. 4.2a) that, in general, similar or higher total extraction yields of 24.8–16.0% are attained employing alkaline solutions at temperatures ranging from 70 to 100 °C. In alkali conditions, the highest extraction yield of 24.80% is achieved employing a 1% NaOH at boiling point for 1 h and solid/liquid ratio of 1:200 (kg/L). Vazquez et al. (2012) used a mixture of 3% NaOH and 1.5% Na_2SO_3, achieving an extraction yield of 21.9% at 100 °C for 1 h and solid/liquid ratio of 1:15 (%w/w). These values are higher than the maximum extraction yield obtained with water (17.7% at 140 °C). The next set of experiments also achieving good extraction yields between 18.9% and 16.0% are obtained employing 0.1% NaOH, 10% NaOH, and a mixture of 5% NaOH–5% Na_2SO_3.

However, the use of alkaline solutions did not improve the TPC extraction yield (Fig. 4.2b) and, more noteworthy, did not improve the extraction selectivity (Fig. 4.2c) for this family of compounds, which means that the extraction of the other type of compounds is being favored by the inclusion of salts into the extracting medium and that the alkaline extracting medium is less selective for the extraction of polyphenols than employing water. Through Fig. 4.2d, it is possible to retrieve that the alkaline extract containing the highest content of TPC in the extract ($18.64\%g_{GAE}/g_{extract}$) is achieved for 2.5% Na_2SO_3 extracting medium, corresponding to almost half the maximum content achieved employing water ($30.6\%g_{GAE}/g_{extract}$).

In accordance, Vazquez et al. (2012) applied a full 2^3 experimental design to evaluate the influence of temperature (90–100 °C) and NaOH (1.5–4.5%) and Na_2SO_3 (0–3%) concentrations on the extraction yield and selectivity for TPC with antioxidant activity and have demonstrated that NaOH concentration increase favors

Table 4.2 Extraction yields of *E. globulus* bark employing water or alkaline solutions at different extracting conditions and respective total phenolic compounds (TPC) content and formaldehyde-condensable tannins (fcT)

Solvent	Extraction conditions	Ext. yield %w/ $w_{\text{dried bark}}$	TPC % g_{GAE}/ g_{ext}	TPC % g_{GAE}/ $g_{\text{dried bark}}$	fcT %w/w_{ext}	References
Water	Room T; 48 h; 1:120 (kg/L)	2.45	30.6	0.75	–	Mota (2011)
	Room T; 24 h; 1:100 (w/v)	1.93	11.53	0.22	–	Santos et al. (2011)
	25 °C; 3.3 h; 1:8 (kg/L)	2.3	10.3	0.24	5.40	Mota et al. (2012)
	82.5 °C; 0.5 h; 1:8 (kg/L)	4.4	20.0	0.88	16.08	Mota et al. (2012)
	82.5 °C; 6 h; 1:8 (kg/L)	5.7	17.2	0.99	18.64	Mota et al. (2012)
	90 °C; 1 h; 1:15 (w/w)	6.8	18.09	1.23	37.6	Vázquez et al. (2008, 2009)
	bp; 3 h; 1:120 (kg/L)	8.14	26.8	2.18	–	Mota (2011)
	140 °C; 3.3 h; 1:8 (kg/L)	17.7	10.9	1.93	6.38	Mota et al. (2012)
	120 °C; 3 h; 0.5:4 (kg/L)	8.0	≈ 25	≈2	–	Pinto et al. (2013)
	140 °C; 3 h; 0.5:4 (kg/L)	14.0	≈ 13	≈1.8	–	Pinto et al. (2013)
0.1% NaOH	bp; 1 h; 1:200 (kg/L)	16.00	6.9	1.10	–	Mota (2011)
	80 °C; 3 h; 0.5:4 (kg/L)	≈ 6	15.6	≈ 1.25	–	Pinto et al. (2013)
	120 °C; 3 h; 0.5:4 (kg/L)	≈ 9	22.6	≈ 2	–	Pinto et al. (2013)
	140 °C; 3 h; 0.5:4 (kg/L)	11.2	20.1	≈ 2.25	–	Pinto et al. (2013)
0.5% NaOH	80 °C; 3 h; 0.5:4 (kg/L)	≈ 8.5	≈ 18.8	≈ 1.6	–	Pinto et al. (2013)
	120 °C; 3 h; 0.5:4 (kg/L)	≈ 12	≈ 25–27.1	≈ 3–3.25	–	Pinto et al. (2013)
	140 °C; 3 h; 0.5:4 (kg/L)	≈15	≈ 25	≈ 3.75	–	Pinto et al. (2013)
1% NaOH	bp; 1 h; 1:200 (kg/L)	24.80	7.7	1.92	–	Mota (2011)
2.5% NaOH	70 °C; 1 h; 1:15 (w/w)	9.8	5.43	0.53	23.1	Vázquez et al. (2009)
	90 °C; 1 h; 1:15 (w/w)	10.8	8.77	0.95	23.5	Vázquez et al. (2009)

(continued)

Table 4.2 (continued)

Solvent	Extraction conditions	Ext. yield %w/ $w_{dried\ bark}$	TPC % g_{GAE}/ g_{ext}	TPC % g_{GAE}/ $g_{dried\ bark}$	fcT %w/w_{ext}	References
10% NaOH	70 °C; 1 h; 1:15 (w/w)	16.4	3.63	0.60	14.3	Vázquez et al. (2009)
	90 °C; 1 h; 1:15 (w/w)	18.9	5.40	1.02	16.5	Vázquez et al. (2009)
1.5% Na$_2$SO$_3$	90 °C; 1 h; 1:15 (w/w)	4.3	21.2	–	–	Vázquez et al. (2012)
	100 °C; 1 h; 1:15 (w/w)	7.1	22.8	–	–	Vázquez et al. (2012)
2.5% Na$_2$SO$_3$	70 °C; 1 h; 1:15 (w/w)	6.8	14.39	0.98	28.9	Vázquez et al. (2009)
	90 °C; 1 h; 1:15 (w/w)	8.55	18.64	1.59	35.5	Vázquez et al. (2008, 2009)
4.5% Na$_2$SO$_3$	90 °C; 1 h; 1:15 (w/w)	8.2	16.8	–	–	Vázquez et al. (2012)
	100 °C; 1 h; 1:15 (w/w)	7.8	17.9	–	–	Vázquez et al. (2012)
10% Na$_2$SO$_3$	70 °C; 1 h; 1:15 (w/w)	6.8	9.15	0.62	9.7	Vázquez et al. (2009)
	90 °C; 1 h; 1:15 (w/w)	10.2	12.48	1.27	15.3	Vázquez et al. (2009)
1.5% NaOH/3% Na$_2$SO$_3$	95 °C; 1 h; 1:15 (w/w)	15.2	14.0	–	–	Vázquez et al. (2012)
2.5% NaOH/2.5% Na$_2$SO$_3$	70 °C; 1 h; 1:15 (w/w)	7.9	7.15	0.56	8.9	Vázquez et al. (2009)
	90 °C; 1 h; 1:15 (w/w)	12.3	9.17	1.13	12.0	Vázquez et al. (2009)
3% NaOH/1.5% Na$_2$SO$_3$	90 °C; 1 h; 1:15 (w/w)	17.5	8.6	–	–	Vázquez et al. (2012)
	100 °C; 1 h; 1:15 (w/w)	21.9	11.8	–	–	Vázquez et al. (2012)
3% NaOH/4.5% Na$_2$SO$_3$	90 °C; 1 h; 1:15 (w/w)	16.2	12.8	–	–	Vázquez et al. (2012)
	100 °C; 1 h; 1:15 (w/w)	14	8.2	–	–	Vázquez et al. (2012)
5% NaOH/5% Na$_2$SO$_3$	70 °C; 1 h; 1:15 (w/w)	13.3	5.02	0.67	5.3	Vázquez et al. (2009)
	90 °C; 1 h; 1:15 (w/w)	16.2	6.56	1.06	5.2	Vázquez et al. (2009)
2.5% Na$_2$CO$_3$/2.5% Na$_2$SO$_3$	90 °C; 1 h; 1:15 (w/w)	10.7	10.30	1.10	14.3	Vázquez et al. (2009)
5% Na$_2$CO$_3$/5% Na$_2$SO$_3$	90 °C; 1 h; 1:15 (w/w)	10.7	5.52	0.59	5.1	Vázquez et al. (2009)

GAE gallic acid equivalent, *AAE* ascorbic acid equivalent, *bp* boiling point

Table 4.3 Proanthocyanidins (Pac) and total carbohydrate content (TC) in *E. globulus* bark aqueous extract

Solvent	Extraction conditions	Pac % $w_{MEE}/w_{dried\ bark}$	Pac % $w_{MEE}/w_{extract}$	TC % $w/w_{extract}$	Tannins % w/w_{bark}
Water	25 °C; 3.3 h; 1:8 (Kg/L)	0.01	0.4	22.4	–
	82.5 °C; 0.5 h; 1:8 (Kg/L)	0.16	3.7	26.7	–
	82.5 °C; 6 h; 1:8 (Kg/L)	0.10	1.8	35.4	–
	140 °C; 3.3 h; 1:8 (Kg/L)	0.16	0.9	50.9	–
	120 °C; 3 h; 0.5:4 (Kg/L)	–	–	–	≈ 1.8–2
	140 °C; 3 h; 0.5:4 (Kg/L)	–	–	–	≈ 2.25
NaOH 0.1%	80 °C; 3 h; 0.5:4 (Kg/L)	–	–	–	1.2
	120 °C; 3 h; 0.5:4 (Kg/L)	–	–	–	2–2.25
	140 °C; 3 h; 0.5:4 (Kg/L)	–	–	–	–
NaOH 0.5%	80 °C; 3 h; 0.5:4 (Kg/L)	–	–	–	–
	120 °C; 3 h; 0.5:4 (Kg/L)	–	–	–	2.5
	140 °C; 3 h; 0.5:4 (Kg/L)	–	–	–	–

Mota et al. (2012), Pinto et al. (2013)
MEE mimosa extract equivalents, Pac, butanol-HCl method, Tannins, hide-powder method

temperatures are applied, then longer times must be applied to optimize the response, or, on the other hand, if higher temperatures are used, the longer times should be avoided in order to prevent degradation of the compounds. For the fcT the extraction time did not exert influence.

In Mota et al. (2012, 2013), it was also demonstrated that the temperature increase favors the co-extraction of carbohydrates (Table 4.3 and Fig. 4.2e) and thus, the extraction performed at 140 °C, having the highest total extraction yield, is the experiment showing the highest carbohydrate content (50.9%w/$w_{extract}$) in the extract and one of the lowest content in TPC (10.9%w/$w_{extract}$).

Among the literature data, the best two sets of extraction conditions selectively extracting TPC are (1) at room temperature for 48 h (30.6% $g_{GAE}/g_{extract}$) and (2) at boiling temperature for 3 h (26.8% $g_{GAE}/g_{extract}$). It is possible to perceive that higher temperatures can selectively enhance the extraction of TPC with less extraction time.

the overall extraction yield but diminishes the selectivity for TPC with antioxidant compounds. Therefore, regarding their experimental design, the best conditions selectively extracting TPC with antioxidant activity correspond to the lowest concentrations of Na_2SO_3 (0%) and NaOH (1.5%) and highest temperature (100 °C). Additionally, the authors have showed that the extracts with higher antioxidant activity corresponded to those having lower molecular weight.

With regard to condensed tannins, through Fig. 4.2d, it is possible to observe that the type of salt and concentration significantly affect the extract potential for wood adhesives. NaOH solutions and 2.5% Na_2SO_3 solutions promote the extraction of condensable tannins, and, on the contrary, employing higher concentrations of Na_2SO_3 and salt combinations of NaOH, Na_2SO_3, and Na_2CO_3 has a negative influence in the content of condensable tannins. In Vázquez et al. (2009) study, it has been demonstrated that the alkalinity conditions decrease the quality of the condensable tannins and the authors have attributed to a probable increase of the extraction of non-condensed tannins. In fact, *E. globulus* bark is known for being rich in hydrolysable tannins (ellagi- and gallotannins), as discussed before in 4.1.

Water extraction performed at 90 °C is the experiment obtaining the highest SN of 37.6, followed by 2.5% Na_2SO_3 aqueous solution and 2.5% NaOH with a SN of 35.5 and 23.5, respectively. The remaining aqueous and alkaline solution studies applied have a SN inferior to 20.

Pinto et al. (2013) also have studied the *E. globulus* bark extractions with water and weak alkaline solutions mainly focused in the extraction of tannins, a particular family of polyphenolic compounds with applications in leather tanning and adhesive formulation (Pizzi 2008). The tannin content of the extracts obtained was evaluated by the hide-powder method (values summarized in Table 4.3), assessing the potential of the bark extracts for leather tanning applications, and TPC content was measured by the Folin-Ciocalteu method (indicated in Table 4.2). Similar to other studies, the authors have observed that the increase of the extraction temperature favored the overall extraction yields for both water and alkaline solutions employed (NaOH 0.1% and 0.5%). Moreover, the NaOH revealed a favorable effect on the TPC extraction allowing to conclude that the alkali solutions improve mass transfer due to bark matrix swelling.

Regardless of the solvent employed, the authors have reached a maximum tannin yield level of approximately 2%wt, corresponding to an overall extraction yield of 10%wt and a TPC extraction yield of 2%wt. After that, extraction conditions leading to higher yields promote the co-extraction of other compounds, consequently decreasing the selectivity for tannins.

Taking into account these findings, the authors did not observe advantages of employing alkaline solutions to obtain an extract rich in tannins suitable for leather tanning applications and selected water as the best extraction medium once it was possible to achieve the maximum extraction selectivity and simultaneously avoid a further neutralization step that alkaline extracts would require. The best conditions selected for extracting tannins were aqueous media at 140 °C for 120 min and an extract containing 2.6 g/L and 2.5 g/L of tannins and TPC, respectively.

The studies pursued with consecutive extractions of fresh bark at the previously selected optimum conditions in an attempt to concentrate the tannin content up to 10 g/L, the typical value employed in retanning applications. This concentration strategy allowed achieving a final concentrated extract containing 8.5 g/L of TPC and 6.5 g/L of tannins, after performing seven extraction cycles. Although the authors observed that this strategy was more effective in concentrating compounds of phenolic nature with no tanning activity, preliminary trials revealed the potential of the final concentrated extract in leather retaining, detailed in Sect. 4.2.3.

4.2.2 Ethanol/Water Extraction: Process Optimization for Phenolic Compounds

Organic solvents are often applied to extract antioxidant compounds from natural sources such as leaves, barks, fruits, etc. Besides the high solubility for polar compounds with important biological properties, they have the clear advantage of solvent recycling by distillation or evaporation.

Regarding *E. globulus* bark, it has been developed studies employing methanol, ethanol, or their mixtures with water. New market trends recommend the use of environmental friendly solvents to recover natural oxidants, and thus, ethanol, water, and their mixtures are a very attractive choice.

In Table 4.4 it is summarized the existing literature data for the extraction studies performed with ethanol aqueous mixtures as the extracting solvents. It is possible to perceive from all ethanol/water mixtures employed that the higher the water content, extraction time, and temperature, the higher the overall extraction and TPC yields are. The best set of conditions achieving the highest overall and TPC extraction yields of 19.4 and 4.0%w/w, respectively, were 140 °C and 6 h.

For the maximum extraction selectivity of TPC with FRAP antioxidant activity, the combination between time and temperature is important. For temperature higher than 82.5 °C, prolonging the extraction time should be avoided since TPC content in the extract decreases. On the contrary, for 25 °C, extending the extraction time from 0.5 h to 6 h, the TPC in the extract increased from 12.2 to 30.8%g_{GAE}. The selectivity for obtaining Pac and fcT from *E. globulus* bark is favored by the presence of ethanol.

Mota et al. (2012) used experimental design to optimize the extraction of polyphenols from *E. globulus* bark employing ethanol water mixtures and understand the influence of the extraction time (30–360 min), temperature (25–140 °C), and ethanol percentage in water (0–80%v/v). The authors have successfully found second-order polynomial equations to describe and predict the extraction yield as well as yield of extraction of TPC, FRAP antioxidant compounds, and Pac and respective contents in the extract as a function of time, temperature, and ethanol % of extraction. Moreover, the potential of the extracts for adhesives formulation and the co-extraction of carbohydrates were also assessed.

Table 4.4 Extraction yields of *E. globulus* bark employing ethanolic mixtures and respective phenolic compounds content (TPC), antioxidant activity (FRAP), and Stiasny number (SN)

Solvent	Extraction conditions	Ext. Yield %w/ $W_{dried\ bark}$	TPC %g_{GAE}/ $g_{dried\ bark}$	TPC %g_{GAE}/ g_{ext}	FRAP $nmol_{AAE}$/ mg_{ext}	PAC %g_{MME}/ g_{ext}	SN	TC %g/ g_{ext}
Ethanol/ water 40:60%v/v	25 °C; 0.5 h; 1:8 (kg/L)	2.8	0.3	12.2	900	7.2	14.7	22.3
	25 °C; 6 h; 1:8 (kg/L)	3.2	1.0	30.8	2100	8.8	17.6	22.5
	140 °C; 0.5 h; 1:8 (kg/L)	8.3	2.0	23.9	1400	9.2	32.6	35.6
	140 °C; 6 h; 1:8 (kg/L)	19.4	4.0	20.7	1200	6.8	33.2	38.9
	83 °C; 3.3 h; 1:8 (kg/L)	5.7	1.8	31.1	2000	14.2	22.3	32.3
Ethanol/ water 52:48%v/v	83 °C; 4.4 h; 1:8 (kg/L)	5.2	1.7	30.4	2160	14.2	40	21.2
Ethanol/ water 80:20%v/v	83 °C; 0.5 h; 1:8 (kg/L)	3.2	0.7	21.0	1500	7.4	20.5	32.5
	83 °C; 6 h; 1:8 (kg/L)	4.3	1.1	25.9	1800	13.9	18.7	44.5
		6.0	1.8	29.6	–	20.4	21.2[a]	33
	25 °C; 3.3 h; 1:8 (kg/L)	2.3	0.4	15.5	0.9	3.2	19.7	27.0
	140 °C; 3.3 h; 1:8 (kg/L)	7.6	2.4	31.4	1.9	9.7	26.3	33.6

Baptista et al. (2015); Mota et al. (2012)
[a]reducing sugars, DNS method

A scheme of the main conclusions is portrayed in Fig. 4.3. Moreover, from the experimental design, the authors have concluded that according to the extraction conditions, different responses could be optimized (Mota et al. 2012). The authors have found that water was the best extracting medium to maximize the overall extraction yield of compounds from *E. globulus* bark and promote the carbohydrate content in the extract. The predicted maximum yield of extraction (21.7%w/$w_{dried\ bark}$) and carbohydrate content in the extract (51.0%w/$w_{extract}$) are obtained for water at maximum extraction time and temperature studied for 360 min and 140 °C, respectively.

The extraction of polyphenolic compounds is favored by ethanol/water mixtures between 40% and 53%. Moreover, the maximum extraction yields of TPC with

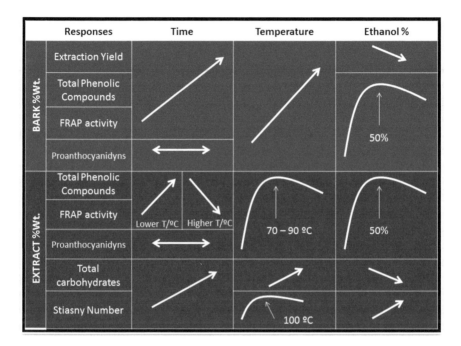

Fig. 4.3 Summary of experimental design study performed by Mota et al. (2012) about the effect of the independent variables time, temperature, and ethanol percentage on the different dependent variables studied

antioxidant activity and Pac are obtained for the maximum extraction time and temperature studied for 360 min and 140 °C. However, with respect to extraction selectivity, it was observed that milder temperatures (70–90 °C) are the most suitable choice. The optimal operating conditions found for TPC content in the extract were found to be for 264 min, 82.5 °C, and 52% ethanol, and similar extraction conditions were obtained for the extraction of FRAP antioxidant compounds (310 min, 72.4 °C, and 52% ethanol). The highest content on Pac in the extract is achieved for 360 min, 87.3 °C, and 53% ethanol.

The condensable tannins with formaldehyde given by the Stiasny number were selectively extracted employing the highest ethanol content studied (80%) for 360 min at 101.5 °C.

4.2.3   Screening for Valuable Applications: Tanning Properties and Biological Activity

Plants contain several compounds that can be considered as potentially safe natural antioxidants to be applied in food and pharmaceutical industries and cosmetics. Some of these compounds have been identified as polyphenols and have been related to provide protection against several diseases due to their antioxidant properties (Moure et al. 2001).

In the previous sections, it has been disclosed that polar *E. globulus* bark extracts are very rich in polyphenols manifesting antioxidant activity (Mota et al. 2012; Vázquez et al. 2009, 2012) and was observed that the extracting conditions such as temperature and extracting medium can influence the antioxidant activity. Studies related with other natural extracts also demonstrated that the extracting conditions affect the biological properties of the extract. Moreover, polymeric polyphenols are more powerful antioxidants than simple monomeric phenolic compounds (Moure et al. 2001); it has already been demonstrated that the condensed and hydrolysable tannins have higher ability at quenching peroxyl radicals than simple phenols (Hagerman et al. 1998), and the higher is the degree of polymerization, the stronger can be the scavenging activity (Yamaguchi et al. 1999) or the inhibition of lipid peroxidation (García-Conesa et al. 1999). Many compounds identified in methanolic/aqueous extracts are known for having important biological activities, which are the case of several flavonoids identified, as, for instance, naringenin, taxifolin, or ellagic acid (García-Niño and Zazueta 2015; Ito et al. 1999; Rice-Evans et al. 1996; Vattem and Shetty 2005; Wilcox et al. 1999). Nevertheless, the synergetic effects of the different polyphenolics extracted and the purity of the compound(s) also can affect the properties of the extract obtained, and thus, the biological activities of the extracts must be evaluated.

Tannins are natural products of relatively high molecular weight with the ability to complex with carbohydrates and proteins. There is a great demand of these compounds for leather manufacturing and medical and pharmaceutical applications, and as adhesives, additives, flotation agents, and cement plasticizers (Pizzi 2008).

Several studies demonstrate that it is possible to obtain tannins condensable with formaldehyde from *E. globulus* bark and that they can be good replacements to phenol in phenol-formaldehyde adhesives in the synthesis of wood adhesives. However, if these extracts are to be considered to be used as wood adhesives, it is important to eliminate the carbohydrate fraction of the extracts.

Mota et al. (2012) study indicated that the presence of ethanol and temperature is a significant parameter to extract tannins condensable with formaldehyde with great potential to be applied in adhesive formulation. The conditions optimizing the extraction of this type of compounds corresponded to 80% ethanol and 100 °C. These extracting conditions were also important to obtain bark extracts with biological activity against breast cancer cells and will be detailed bellow.

Tanning Properties
Leather tannin refers to the sequence of steps involved in converting animal hides or skins into finished leather. Initially, the hides are submitted to a series of steps aiming to clean and prepare it for tannin. Then the tanning agent combines with the collagen molecules, converting the protein of the raw hide or skin into a more stable and resistant material suitable for several end-applications. Additionally, leather can be retanned, dyed, or fatliquored according to the type of product to be obtained (Thorstensen 1993).

The acid salts of trivalent chromium are the most common tanning agents employed followed by the forest-derived natural vegetable tannins, both accounting for about 90% of the leather manufactured today (Covington 1997; Pizzi 2008). The first product is mainly used to manufacture soft leathers suitable for shoe uppers and for leather bags, and the latter to obtain more heavy, rigid, and hard leathers suitable for shoe soles, saddles, belts, and other products subject to intensive wear. Vegetable tannins and synthetic tannins are applied in either leather tanning or retanning processes. Other products employed to produce specific types of leather are aldehydes (e.g., formaldehyde or glutaraldehyde) and sulfonated synthetic polymers (e.g., acid phenol-formaldehyde novolac-type resins and other synthetic resins and compounds such as acrylics, oxazolidines, and aminoplastic resin).

Although, chromium salts have the enormous drawback of pollution problems and legislation is becoming more restricted regarding the application of this type of agents, the quest to find other suitable replacements is still undergoing (Pizzi 2008). Additionally, traditional leather tanning processes are being improved to be more environmentally friendly and to promote chemical recycling (Covington 1997).

On the other hand, leather produced with vegetable tannins has been limited due to the darkening problems when the product is exposed to light or shrinking problems occurred at lower temperatures. This problem is being overcome with the combination with aminoplast melamine-urea-formaldehyde (MUF) resins to reduce photooxidation and achieve leather shrinkage temperatures matching the chromium salt leathers (Pizzi 2008).

Vegetable tannins such as mimosa, chestnut tree, and quebracho extracts and synthetic tannins have been typically used for leather tanning and retanning (Covington 1997). Typically employed vegetal extracts in retanning procedures contain about 70% tannins and a concentration of about 10 g/L.

Pinto et al. (2013) demonstrated that concentrated aqueous *E. globulus* bark extract has potential to be applied in leather retanning with a comparable performance to chestnut extracts. The authors have applied a concentrated bark aqueous extract containing 6.5 g/L of tannins ($0.74\%w_{tannin}/w_{leather}$) under similar conditions employed with commercial tannins, and comparable tear and tensile strengths and elongation at break to chestnut extract were obtained: 120 N tear strength, ≈ 20 Nmm^2 tensile strength, and 70% elongation at break. *E. globulus* bark extract astringency was comparable to that of the chestnut extract, and its application in leather retanning has resulted in leather with high fullness, large grain, and less break pipiness, well suitable for the production of leather articles like box calf and nubuck.

Mota et al. (2012) preliminary screening for biological activity of *E. globulus* bark extracts demonstrated the potential of aqueous/ethanol extracts to reduce human breast cancer cell proliferation.

In Table 4.5 it is summarized for several bark extracts obtained by the authors the IC50 values, i.e., the extract concentration necessary to reduce in 50% the proliferation of breast cancer cells MDA-MB-231. The authors observed that water was not a good extracting medium for obtaining compounds with activity against proliferation of the studied carcinoma cells and that ethanol incorporation was essential to obtain extracts manifesting desired antiproliferative effect. In fact, a

Table 4.5 IC_{50} values obtained for *E. globulus* bark extracts regarding their activity in regulating MDA-MB-231 human cell proliferation when incubated with different concentrations for 48 h

Extraction conditions			
Time (h)	Temperature (°C)	Ethanol %	IC_{50} (µg/mL)
3.25	25	0	No effect
3.25	140	0	No effect
0.5	25	40	No effect
6	140	40	266.7 ± 18.7
4.4	82.5	52	176.4 ± 15.6
3.25	25	80	244.8 ± 28.3
3.25	140	80	220.1 ± 19.6
6	82.5	80	91.9 ± 9.0

Mota et al. (2012)

positive correlation between biological activity tested and the extract content in tannins condensable with formaldehyde (given by the Stiasny number) was observed. The aqueous extracts were the ones having the lowest Stiasny number. On the other hand, the highest composition of ethanol employed in bark extractions of 80% was the one having the highest Stiasny number value and the lowest IC_{50} concentration. The second best extract manifesting biological activity was obtained with 52% ethanol, having the second highest value of SN. In the same study, it was observed that moderate temperatures employed in bark extraction benefited the biological activity of the extracts obtained.

4.3 Fractionation of Ethanolic Extracts from *Eucalyptus globulus* Bark

4.3.1 Membrane Processing

Membrane processing is a promising technology to refine natural extracts containing phenolic compounds with bioactive properties once it is considered a green process that offers the possibility of fractionating or concentrating the desired solutes present in the extracts without resourcing to harmful chemicals that can compromise the end-uses of the final solution. Moreover, this technology has the additional advantage of operating at mild temperatures, avoiding thermal degradation of the mixture.

Membrane separation is a well-established technology and commonly applied in food processing industry and in the treatment of wastewaters (Daufin et al. 2001). Ultrafiltration and nanofiltration are the more commonly studied processes to concentrate or fractionate the polyphenolic fraction coming from different processing streams or extracts (fruit juices, winery sludge, industrial waste liquors, grape seed extracts, cork processing wastewaters, among others) (Cassano et al. 2008; Díaz-Reinoso et al. 2009, 2010; Minhalma and de Pinho 2001).

In the particular case of *E. globulus* bark, this raw material is an interesting source for bioactive polyphenols, and a membrane fractionation can be important to concentrate or fractionate the obtained extracts. Previously, it was disclosed that the ethanolic solutions can selectively extract polyphenolic compounds from *E. globulus* bark, and a preliminary study on membrane processing of these extracts became essential in order to understand the viability of applying this technology to improve the extracts obtained in terms of promoting the enrichment of polyphenols in detriment of other compounds, such as total carbohydrates. Several operating conditions influence membrane performance such as temperature, pH, feed concentration, chemical composition of the membrane, pore size, feed flowrate, and transmembrane pressure. Ethanol will also affect membrane productivity and rejection coefficients once it changes solvent-solute-membrane interactions, thus affecting the transport mechanism and solute retentions (Geens et al. 2005). Several works have demonstrated that rejection coefficients to a certain solute can be significantly altered by the presence of ethanol and according to the composition of the membrane, the effect of ethanol can be significantly different (Geens et al. 2005; Pinto et al. 2014a).

Two promising *E. globulus* bark extracts were processed by membrane technology: a first one obtained with ethanol/water (80:20 v/v) solution having the best performance against breast cancer cells and having the highest content in formaldehyde-condensable tannins (a particular family of polyphenols) and a second one obtained with ethanol/water (52:48 v/v) solution having the highest content in phenolic compounds in the extract quantified by the Folin-Ciocalteu method. The membranes employed in these studies are summarized in Table 4.6, and the permeabilities of these membranes to water, 52% ethanol, and 80% ethanol are displayed in Table 4.7.

Different study approaches were conducted with these two extracts and are summarized in Fig. 4.4. With the ethanol/water (80:20 v/v) extract, the effect of transmembrane pressure was evaluated in several membranes with different molecular weight cutoff and compositions. The membrane productivity and rejection toward total solids and total phenolic compounds were assessed for each membrane and

Table 4.6 Characteristics of UF/NF membranes employed in *E. globulus* bark extract membrane processing studies

Membrane	Abbrev.	Rejection/MWCO[a]	Type	Producer	Composition
EW	EW60	60,000	UF	GE Osmonics	Polysulfone
JW	JW30	30,000	UF	GE Osmonics	Polyvinylidene
Pleiade	P5	5000	UF	Orelis Environmental	Polyethersulfone
GE	GE1	1000	UF	GE Osmonics	Polyamide composite
90801	–	~ 50% NaCl 90% colorant (350 Da) in ethanol	NF	Solsep	Polyamide derivative

[a]*MWCO* molecular weight cutoff (Da)

Table 4.7 Characteristics of UF/NF membranes employed in *E. globulus* bark extract membrane processing studies

Membrane	Permeability (L m^{-2} h^{-1} bar^{-1})		
	Water	Ethanol/water (52:48, % v/v)	Ethanol/water (80:20, %v/v)
EW60	191.3	≈48.5 39.6	57.2
JW30	91.2	47.6	35.5
P5	72.4	36.4	25.0
GE1	3. 9	1.24	1.5–1.7
90801	–	5.8	1.8

Pinto et al. (2014a)

transmembrane pressure studied. Fouling and cake buildup were evaluated in order to understand permeate flux decrease over processing time with the extract and to confirm the possibility of reutilization of the membranes, two important aspects vital to industrial implementation of a membrane separation process.

Regarding the ethanol/water (52:48 v/v), this bark extract was treated by different ultra- and nanofiltration membranes aiming to understand the effect of the type of membrane in the concentration of the polyphenolic compounds determined by the Folin-Ciocalteu method. The composition of specific families of phenolic compounds (proanthocyanidins and tannins) and total carbohydrates was also assessed. Later, diafiltration and purification by adsorption were employed in an attempt to enrich the extract in the desired phenolic compounds. It is expected that in diafiltration process, the retentate stream will be further depleted in the lower molecular weight compounds comparatively to the UF process (Fikar et al. 2010). This process has already been suggested as a plausible methodology for fractionation or intensification of the polyphenolic content present in the feed solution (Acosta et al. 2014; Russo 2007; Santamaría et al. 2002).

4.3.1.1 Resistance and Cake Buildup Analysis in the Ultrafiltration of Ethanol/Water Extract (80:20 v/v)

Ultrafiltration of *E. globulus* bark extract obtained with ethanol/water (80:20 v/v) solution was studied by Baptista et al. (2015) employing different MWCO membranes, and compositions were used: GE1, P5, JW30, and EW60 (composition and MWCO are summarized in Table 4.7).

The study was performed at 35 °C and 4.5 L min^{-1} for transmembrane pressures of 3, 5, and 8 bar, and the respective permeate flux evolution with time is portrayed in Fig. 4.5. The experiments were conducted in full recycle mode and in sequence starting with the lowest pressure, and, therefore, typical flux reduction curve (Song 1998) is only observed at 3 bar, the first pressure tested: an initial accentuated drop of flux (stage 1), a long-term gradual decrease (stage 2), and a stationary state flux (stage 3).

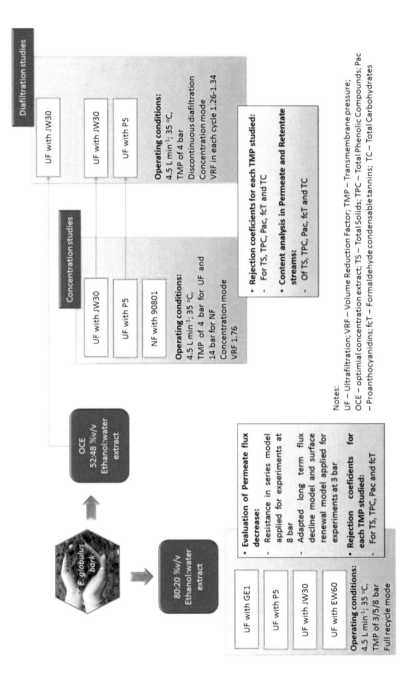

Fig. 4.4 Overview of the membrane separation studies performed with ethanolic extracts of *E. globulus*

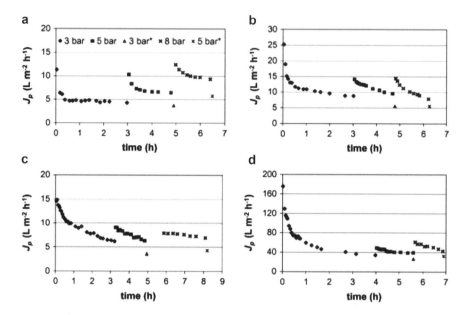

Fig. 4.5 Evolution of the permeate flux (J_P) with time for (**a**) GE1, (**b**) P5, (**c**) JW30, and (**d**) EW60, operated at different transmembrane pressures sequentially, in full recycling mode, 35 °C, and a feed flow of 4.5 L min⁻¹. *pressure applied after operation at 5 or 8 bar. (Reprinted from Baptista et al. (2015), Copyright (2015), with permission from Elsevier)

Table 4.8 Permeate flux of the solvent 80:20 v/v ethanol/water (S) and respective *E. globulus* bark extract (E) in the stationary state for different transmembrane pressures

TMP bar	J (L m⁻² h⁻¹)							
	GE1		P5		JW30		EW60	
	S	E	S	E	S	E	St	E
3	5.4	4.3	97.7	6.2	114.6	8.8	237.4	34.1
5	8.8	6.4	152.6	6.3	185.0	9.5	366.0	30.1
8	13.9	9.4	227.4	6.9	295.0	8.9	529.5	42.5

Baptista (2013), Baptista et al. (2015)

The membrane permeability of ethanol/water (80:20 v/v) and the correspondent *E. globulus* bark extract reached at steady state are summarized in Table 4.8. The permeate flux not only depends on the MWCO of the membrane but also is affected by the chemical nature of the membrane, therefore explaining the fact that although GE1, P5, and JW membranes have significantly different MWCO, the permeate flux to ethanol/water (80:20 v/v) is very similar. When changing from the solvent solution to the real extract, the permeate flux has decreased slightly for GE membrane (20–33% decrease) and more severely for P5, JW30, and EW60 membranes (86–76%), demonstrating that fouling formation was more relevant for these last three membranes.

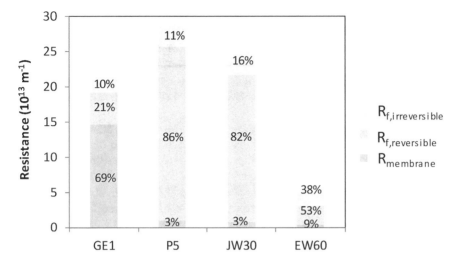

Fig. 4.6 Resistances determined for each membrane processing of the bark ethanol/water extract (80:20 v/v) at 8 bar by applying resistance in series model. (Adapted from Baptista et al. (2015), Copyright (2015), with permission from Elsevier)

Baptista et al. (2015) study revealed that the transmembrane pressure (TMP) increase led to higher formation of fouling since after conducting experiments at 5 and 8 bar, the permeate flux recorded for the previous TMP studied of 3 and 5 bar, respectively, was lower (Fig. 4.5). This flux decrease observed was more accentuated for P5 and JW30 membrane processing, demonstrating that these membranes were the ones more affected by fouling formation. Fouling resistances experimentally determined for each TMP studied and models applied by the authors to describe flux decrease at 3 bar (adapted long-term flux decline model and surface renewal) also support these findings.

In Fig. 4.6 it is shown the contribution of each resistance in the UF of the bark extract obtained by Baptista et al. (2015) at 8 bar with different membranes where it is observed that resistance caused by fouling formation is very relevant for P5 and JW30 membranes, representing 97–98% of the total resistance observed. Regarding EW60 membrane processing, resistance caused by fouling accounts for 91% of the total resistance observed, and the least affected membrane by fouling is the GE membrane with about 31% of the total resistance caused by fouling. Moreover, the EW60 membrane had a significant contribution of the irreversible fouling, constituting a disadvantage due to the fact that processing with this membrane will require chemical cleaning in order to allow successive membrane processing and could affect the long-term use of the membrane.

In membrane processing understanding the key factors responsible for permeate flux decline is important to assess the viability of developing an industrial process. Permeate flux decline can be described considering three distinct phases (Song 1998): a first stage with an accentuated permeate flux decrease due to pore blocking, a second stage where permeate flux gradually decreases due to gel polarization and cake layer buildup on the membrane surface, and a third stage where cake resistance increases while the cake gets thicker until reaching a steady-state plateau.

In order to understand permeate flux decline during membrane processing with the ethanolic extract, Baptista et al. (2015) implemented the surface renewal model (Hasan et al. 2013) and the adapted long-term flux decline (De et al. 1997). Both mathematical models successfully described flux decline for processing the bark extract at 3 bar.

The surface renewal model is based upon the cake filtration theory and the surface renewal concept and considers that cake formation is the dominant fouling mechanism responsible for flux decline and the liquid elements are continuously brought into contact from the bulk liquid to the membrane surface and remain for a definite time before returning again to the bulk liquid. This method allows the prediction of the rate of renewal of liquid elements at the membrane surface and reflects the easiness to remove the deposited material; the higher the rate, the easier it is to remove the deposited material; and thus, this is directly reflected on the recovery of the initial membrane permeability after cleaning. The highest rates of renewal of liquid elements corresponded to the membranes JW30 and GE1, and, in accordance, these membranes were the easiest to clean, and more than 90% of the initial permeability was accomplished within one to two cycles of chemical cleaning with NaOH 0.1 M in ethanol/water (80:20 v/v). On the other hand, EW60 and P5 were the membranes with the lowest rates, where two to four chemical cleaning cycles were not sufficient to recover the initial permeate flux of the membranes. Moreover, the rate was very similar for EW60 and P5 membranes, allowing to conclude that composition of the membranes is detriment to the performance of the membrane.

This model also helps understanding the cake buildup on the membrane surface during processing at 3 bar. EW60 was the membrane process showing the highest growth of cake layer, followed by JW30 and GE1 exhibiting a similar propensity, and P5 was the one with less mass accumulated of cake on the surface.

The adapted long-term flux decline model considers two distinct time domains, an initial short term where flux decrease is very fast and is mainly due to osmotic pressure and a long term where flux decrease is slower mainly due to gradual growth of the polarized layer until reaching a steady state. This model allowed confirming that P5 and EW60 are the membranes exhibiting lower polarization and cake layer growth rate and thus, reversible fouling growth is gradual explaining the fact that processing with these membranes took longer time to reach steady state. On the other hand, GE membrane is the one exhibiting the highest growth velocity of polarization concentration and cake buildup, and therefore, the steady state is reached faster. This model allowed estimating the total resistance to flux being the values obtained in very close agreement with the experimental ones obtained. Moreover, it demonstrated that P5 membrane has the additional advantage of being the least susceptible membrane to reversible fouling.

Membrane productivity is not the only variable to consider when evaluating a membrane separation process, but also the rejection toward the solutes of interest is very important in order to obtain a desired concentration or fractionation process. The rejection coefficients of solutes are not directly related with the MWCO of the

Table 4.9 Rejection coefficients to TS, TPC, Pac, and fcT obtained with membrane processing at different TMP pressures of 80:20% v/v *E. globulus* bark extract bark with different membranes

	3 bar				5 bar				8 bar			
	GE	P5	JW30	EW60	GE	P5	JW30	EW60	GE	P5	JW30	EW60
TS	43	48	51	15	52	55	60	13	53	63	57	14
TPC	49	62	61	14	58	66	66	5	61	73	71	4
Pac	66	87	82	30	76	88	87	16	80	93	92	14
fcT	75	73	–	28	80	83	–	12	–	–	–	18

Conditions: 35 °C, feed flow rate 4.5 L min^{-1}. Operation took place in full recycle mode, and permeate samples were collected after the permeate flux stabilized
Baptista et al. (2015)

membranes; other factors such as membrane composition, possible interaction between solutes, and membrane surface or propensity for cake layer formation will affect rejection coefficients observed.

In Table 4.9 it is summarized the rejection coefficients obtained for total non-volatile solids (TS), total phenolic compounds (TPC), proanthocyanidins (Pac), and formaldehyde-condensable tannins (fcT) with the processing of the ethanolic bark extract. The TPC content is found by the Folin-Ciocalteu method, Pac by the acid butanol method, and fcT by the Stiasny method.

In general, rejection coefficients have increased with pressure for all tested membranes except for the highest cutoff membrane of 60 kDa, the membrane showing the lowest rejection coefficients. The increase of rejection coefficients with the TMP can be explained by the higher compression of the polarization layer. P5 and JW30 were the membranes where the rejection coefficients were more affected by TMP and, in accordance, were also the membranes exhibiting higher fouling formation.

The membranes showing the highest rejection coefficients toward TS, TPC, and Pac were P5 and JW that did not correspond to the lowest cutoff membrane employed (GE 1 kDa) in the study, revealing that the composition of the membranes had a relevant role in the separation.

Considering that the main goal of the membrane separation process is to retain the highest amounts of TPC and Pac with the highest membrane productivities, Baptista et al. (2015) have concluded that the best operating pressure is of 3 bar for the membrane EW and 5 bar for the remaining membranes studied. Since P5 was the membrane more difficult to clean, JW membrane seems the most promising one to be applied to treat the ethanol/water (80:20 v/v) *E. globulus* bark extract.

4.3.1.2 Application of Ultrafiltration and Nanofiltration to Ethanol/Water Extract (52:48 v/v)

The membrane concentration study of the *E. globulus* bark ethanol/water extract (52:48 v/v) was performed by Pinto et al. (2014a) aiming to increase the content of the extract in polyphenols. The authors selected three commercial membranes, two UF membranes, and one NF membrane to proceed with the bark extract

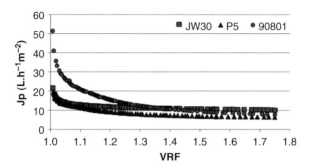

Fig. 4.7 Instantaneous permeate flux (Jp) with time for the 52:48% v/v ethanol/water *E. globulus* bark extract. Conditions: concentration mode, 35 °C, feed flow 4.5 L min^{-1}, TMP 4 bar for ultra-filtration JW30 and P5 and 14 bar for nanofiltration 90,801; operation time for VRF of 1.76, 4.3 h, JW30; 7.4 h, P5; 5.6 h, 90801. (Reprinted from Pinto et al. (2014a), Copyright (2014), with permission from Elsevier)

concentration study, and the final product composition was characterized regarding TS, TPC, Pac, fcT, and total carbohydrates (TC) obtained by methanolysis.

Concentration experiments were carried out up to a volume reduction factor of 1.76, in batch mode with retentate recycling and permeate removal at constant TMP, feed flow, and temperature. The membranes selected were the UF JW30 and P5 and the NF 90801 (composition and MWCO are summarized in Table 4.7).

In Fig. 4.7 it is displayed the permeate flux evolution with time of processing. Among the studied membranes, JW30 had the best permeate flux performance. P5 was the membrane with the lowest permeate flux which was reflected in the longer operating time among the other membranes. NF 90801 was the membrane with the highest permeate decline, and, after 2.5 h of operating time, the permeate flux was lower than the one observed for JW membrane.

After each optimized condition of extraction (OCE) processing, cycles of membrane cleaning were performed with alkaline solutions; 0.1 M prepared in ethanol/water (52:48 v/v) solution has shown that more than 80% of the initial permeate flux was restored within one to two cleaning cycles for the UF membranes and three cycles for the NF membrane. This is a good indicator that irreversible fouling is not significant for all the three membrane processing and that a membrane separation process employing these membranes can be envisaged.

Membrane apparent rejections concerning TS, TPC, and TC assayed by Pinto et al. (2014a) for each membrane and are summarized in Table 4.10. 90801 was the membrane with the highest rejection coefficients, and JW30 being the membrane with the largest cutoff, the apparent rejection coefficients obtained were lower than the values registered for the other two.

In Fig. 4.8 the composition of each final retentate regarding phenolic compounds quantified as TPC, Pac, and fcT is displayed along with the TC content obtained.

The TPC content was highest in retentates obtained with the membranes P5 and 90,801 indicating that the lower MWCO membranes retain more solutes carrying

Table 4.10 Rejection coefficients to TS, TPC, and TC obtained with membrane processing of 52:48% v/v *E. globulus* bark extract with membranes JW30, P5, and 90801

%Rej.	JW30	P5	90801
TS	58	77	78
TPC	75	85	92
TC	12	17	15

Conditions: 35 °C, feed flow rate 4.5 L min⁻¹, TMP 4 bar for UF (JW30 and P5) and 14 bar for NF (90801). Operation was in concentration mode until reaching a VRF of 1.76
Pinto et al. (2014a)

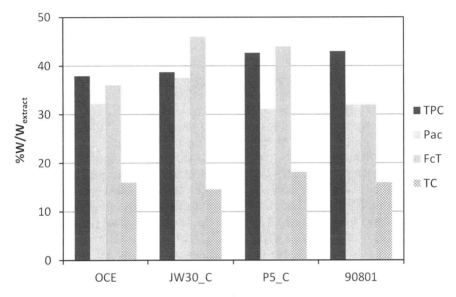

Fig. 4.8 Composition of the feed (OCE) and respective retentates produced with membranes JW30, P5, and 90,801. The values are presented as % weight/dry weight of the OC extract (the feed) or dry weight of the retentate obtained for each membrane. Conditions: concentration mode (VRF 1.76), 35 °C, feed flowrate 4.5 L min⁻¹, TMP 4 bar for UF JW30 and P5 membranes and 14 bar for NF 90801 membrane. (Adapted from Pinto et al. (2014a), Copyright (2014), with permission from Elsevier)

phenolic hydroxyl groups, probably associated with low molecular weight phenolic compounds present in the extract. On the other hand, JW30 membrane was more effective to retain the particular family of compounds Pac (mainly flavonoid oligomers) and fcT (mono- and biflavonoids and some oligomers). The content in TPC, Pac, and fcT in the final retentate of JW30 membrane accomplished by Pinto et al. (2014a) revealed that this membrane was more selective to retain phenolic compounds in detriment of TC. The final bark extract, treated with JW30 membrane, was enriched by 2%, 16%, and 28% of TPC, Pac, and fcT, respectively, and lost 10% TC.

Additionally, the authors have also detailed TC composition in the initial feed on each retentate and permeate streams obtained and observed a modification of the relative composition of the carbohydrate fraction due to a preferential permeation of glucose and galacturonic acid-containing oligo−/poly-saccharides and higher rejections by the membranes to rhamnose and arabinose moieties, probably linked with the polyphenolic compounds (Pinto et al. 2014a).

4.3.2   Diafiltration and Adsorption for Purification and Concentration of Polyphenols

Discontinuous diafiltration (DF) studies were conducted by Pinto et al. (2017) aiming to further increase the selectivity for polyphenols observed in the membrane concentration experiments of the ethanol/water (52:48 v/v) extract promoted by the washing effect of the DF process. This procedure has the additional advantage of reducing the ethanol content of the extract to a value below 12% and allows a posterior adsorption process step to further purify the bark extract.

The three diafiltration processing approaches studied are outlined in Fig. 4.4: the direct diafiltration of the OCE (D_OCE) with membrane JW30 and pre-concentration of the OCE up to a VRF of 1.76 and diafiltration of the retentate with membranes JW30 and P5 (D_JW30_C and D_P5_C). Six discontinuous diafiltration cycles were performed at 35 °C and fixed TPM of 4 bar. In each cycle, the permeate stream was collected until achieving a VRF of 1.26/1.34 and replaced by the same volume with water. In the case of the pre-concentrated approaches, the initial volume was made up with water after concentrating the OCE.

In Fig. 4.9 it is shown the permeate fluxes observed by Pinto et al. (2017) during the diafiltration studies along with the evolution of the volume reduction factor for each stage. The diafiltration with P5 membrane was the approach exhibiting the lowest permeate flux in all cycles. Regarding processing with the membrane JW30, it was observed that the permeate flux was somewhat higher when diafiltration was performed after a pre-concentration step probably explained by the removal of some compounds during the pre-concentration step that generate fouling.

In Fig. 4.10 it is depicted the OCE and final retentate composition of the three diafiltration approaches studied by Pinto et al. (2017) concerning TPC, Pac, FcT, and TC. All diafiltration processes studied allowed the enrichment of the polyphenolic content (TPC, Pac, and FcT). Diafiltration of the pre-concentrated OCE extract was the approach leading to higher concentration of the phenolic compounds present in the bark extract, in particular for Pac and fcT, where an increment of 64% and 55% was observed, respectively. Moreover, the authors have observed that TC content did not vary much for both JW membrane processings, whereas for the P5 membrane processing, they were somewhat enriched. Similar to what was observed in the concentration studies, a preferential retention of arabinose and rhamnose was observed for all approaches studied.

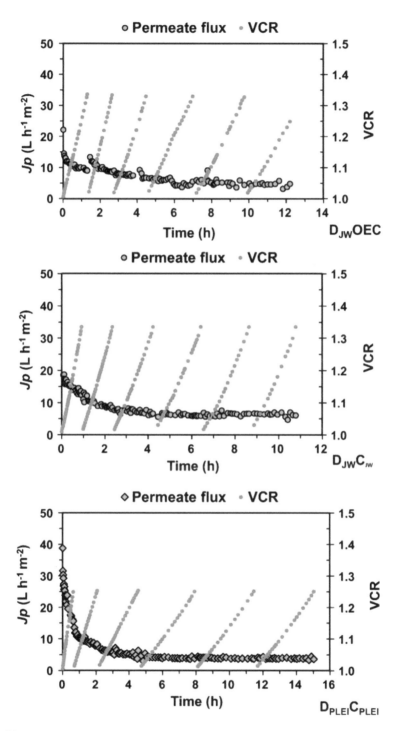

Fig. 4.9 Instantaneous permeate flux and VRF evolution for each cycle during the DF of the ethanol/water (52:48 v/v) *E. globulus bark* extract: (**a**) D_JW; (**b**) D_JW_C; (**c**) D_P5_C. Conditions: 35 °C, feed flowrate 4.5 L min^{-1}, and TMP 4 bar. Water was the replacement solvent. (Reprinted from Pinto et al. (2017), Copyright (2017), with permission from Elsevier)

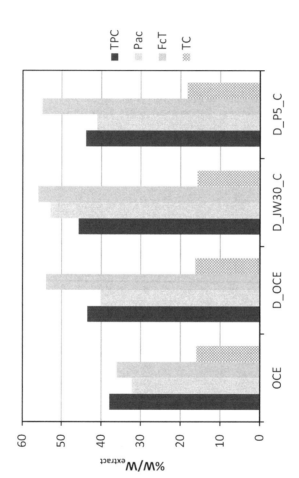

Fig. 4.10 Composition of the feed (OCE) and final retentates obtained after the six diafiltration cycles performed at 35 °C, feed flowrate 4.5 L min⁻¹, and TMP 4 bar for the three approaches studied (as indicated in Fig. 4.4): OCE is the optimum conditions extract, D_OCE corresponds to the direct diafiltration of OCE, and D_JW30_C and D_P5_C refer to the diafiltration of the pre-concentrated OCE by JW30 and P5 membranes, respectively. The values are presented as % weight/dry weight of the OCE extract (the feed) or dry weight of each retentate

Table 4.11 Composition of the streams resulting from processing *E. globulus* bark extract

	OCE	D_JW_C	Enriched fraction
TS, g/L	7.9	5.1	18.7
TPC, g/L	3.0	2.3	12.4
TPC, %w/w	37.7	45.7	66.4
Pac, %w/w	32.3	52.9	55.1
TC, %w/w	16.1	15.8	8.7

Pinto et al. (2017)

Considering the results, the authors have selected the diafiltration of the pre-concentrated OCE with JW membrane to pursue with further refinement by adsorption onto nonpolar resin. After loading this stream onto a bed packed with SP700 resin, the adsorbed compounds were readily recovered with few bed volumes of ethanol/water solution (95:5 v/v) allowing to enrich about 70–75% times the TPC and Pac content in the extract.

In Table 4.11 it is summarized the composition of the overall streams of the separation and purification process employed. The adsorption/desorption step revealed to be a selective process to phenolic compounds in detriment of TC, and it was obtained a final enriched solution containing 66% of TPC, 55% of Pac, and only 9% of TC.

The sequential process of concentration/diafiltration and adsorption/desorption was successfully demonstrated by Pinto et al. (2017), reaching to a final solution of 12 g/L and a content of 67% on dried extract basis. The recovery of the compounds with an ethanolic solution is advantageous considering that it allows energy saving in the drying step. In this work the authors have successfully demonstrated that a membrane separation and adsorption processes can be easily integrated and constitute a promising approach for integration in biorefining processes for bark valorization, attending several green engineering and green chemistry principles. It has been suggested by Pinto et al. (2017) an integrated extraction, separation, and purification process aiming the valorization of *E. globulus* bark as an important source for chemicals (TPC, sugars), to be implemented before burning bark for energy (Fig. 4.11).

Further investigation should be conducted next regarding physical and chemical characterization of the *E. globulus* bark extract obtained in order to make a screening for possible effects and qualities of the extract and thus identify the most promising application(s). Its efficacy, bioavailability, and chemical and physical behavior in several matrices during formulations and potential toxicity should be clearly detailed in order to formulate a safe and marketable product.

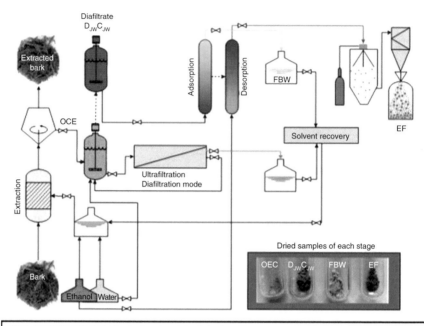

Fig. 4.11 Diagram of the process for separation and recovery of polyphenols from *Eucalyptus globulus* bark encompassing extraction with ethanol/water solution, ultrafiltration, and adsorption onto microporous resin. (Reprinted from Pinto et al. (2017), Copyright (2017), with permission from Elsevier)

References

Acosta O, Vaillant F, Pérez AM, Dornier M (2014) Potential of ultrafiltration for separation and purification of ellagitannins in blackberry (Rubus adenotrichus Schltdl)juice. Sep Purif Technol 125:120–125

Balasundram N, Sundram K, Samman S (2006) Phenolic compounds in plants and agri-industrial by-products: antioxidant activity, occurrence, and potential uses. Food Chem 99:191–203

Baptista EA (2013) Ultrafiltração de extrato de casca de *Eucalyptus globulus* para recuperação de compostos polifenólicos. Master thesis, Faculty of Engineering of University of Porto

Baptista EA, Pinto PCR, Mota IF, Loureiro JM, Rodrigues AE (2015) Ultrafiltration of ethanol/water extract of *Eucalyptus globulus* bark: resistance and cake build up analysis. Sep Purif Technol 144:256–266

Bravo L (1998) Polyphenols: chemistry, dietary sources, metabolism, and nutritional significance. Nutr Rev 56:317–333

Cadahía E, Conde E, deSimon BF, GarciaVallejo MC (1997) Tannin Composition of *Eucalyptus camaldulensis, E. globulus* and *E. rudis.* Part II. Bark. Holzforschung - Int J Biol Chem Phys Technol Wood 51:125–129

Cassano A, Donato L, Conidi C, Drioli E (2008) Recovery of bioactive compounds in kiwifruit juice by ultrafiltration. Innovative Food Sci Emerg Technol 9:556–562

Conde E, Cadahía E, García-Vallejo MC, Tomás-Barberán F (1995) Low molecular weight polyphenols in wood and bark of *Eucalyptus globulus.* Wood Fiber Sci 27:379–383

Conde E, Cadahia E, Diez-Barra R, García-Vallejo MC (1996) Polyphenolic composition of bark extracts from *Eucalyptus camaldulensis, E. globulus* and *E. rudis.* Eur J Wood Wood Prod 54:175–181

Covington AD (1997) Modern tanning chemistry. Chem Soc Rev 26:111–126

Daufin G, Escudier JP, Carrère H, Bérot S, Fillaudeau L, Decloux M (2001) Recent and emerging applications of membrane processes in the food and dairy industry. Food Bioprod Process 79:89–102

De S, Dias JM, Bhattacharya PK (1997) Short and long term flux decline analysis in ultrafiltration. Chem Eng Commun 159:67–89

Díaz-Reinoso B, Moure A, Domínguez H, Parajó JC (2009) Ultra- and nanofiltration of aqueous extracts from distilled fermented grape pomace. J Food Eng 91:587–593

Díaz-Reinoso B, González-López N, Moure A, Domínguez H, Parajó JC (2010) Recovery of antioxidants from industrial waste liquors using membranes and polymeric resins. J Food Eng 96:127–133

Eyles A, Davies N, Mohammed C (2004) Traumatic oil glands induced by pruning in the wound-associated phloem of *Eucalyptus globulus*: chemistry and histology. Trees Struct Funct 18:204–210

Fechtal M, Riedl B (1991) Analyse des Extraits Tannants des Écorces des Eucalyptus après Hydrolyse Acide par la Chromatographie en Phase Gazeuse Couplée avec la Spectrométrie de Masse (GC-MS). Holzforschung - Int J Biol Chem Phys Technol Wood 45:269–273

Fikar M, Kovács Z, Czermak P (2010) Dynamic optimization of batch diafiltration processes. J Membr Sci 355:168–174

García-Conesa MT, Wilson PD, Plumb GW, Ralph J, Williamson G (1999) Antioxidant properties of 4,40-dihydroxy-3,30-dimethoxy-beta, beta-bicinnamic acid (8-8-diferulic acid, non cyclic form). J Sci Food Agric 79:379–384

García-Niño WR, Zazueta C (2015) Ellagic acid: pharmacological activities and molecular mechanisms involved in liver protection. Pharmacol Res 97:84–103

Geens J, Peeters K, Van der Bruggen B, Vandecasteele C (2005) Polymeric nanofiltration of binary water–alcohol mixtures: influence of feed composition and membrane properties on permeability and rejection. J Membr Sci 255:255–264

Hagerman A, Riedl K, Jones G, Sovik K, Ritchard N, Hartzfeld P, Riechel T (1998) High molecular weight plant polyphenolics (tannins) as biological antioxidants. J Agric Food Chem 46:1887–1892

Harborne J, Williams C (2000) Advances in flavonoid research since 1992. Phytochemistry 55:481–504

Hasan A, Peluso CR, Hull TS, Fieschko J, Chatterjee SG (2013) A surface-renewal model of cross-flow microfiltration. Braz J Chem Eng 30:167–186

Hertog M, Hollman P, Venema D (1992) Optimization of a quantitative HPLC determination of potentially anticarcinogenic flavonoids in vegetables and fruits. J Agric Food Chem 40:1591–1598

Ito H et al (1999) Anti-tumor promoting activity of polyphenols from Cowania mexicana and Coleogyne ramosissima. Cancer Lett 143:5–13

Kim J-P, Lee I-K, Yun B-S, Chung S-H, Shim G-S, Koshino H, Yoo I-D (2001) Ellagic acid rhamnosides from the stem bark of *Eucalyptus globulus.* Phytochemistry 57:587–591

Maksimovic Z, Malencic Đ, Kovacevic N (2005) Polyphenol contents and antioxidant activity of Maydis stigma extracts. Bioresour Technol 96:873–877

Marston A, Hostettmann K (2006) 1. separation and quantification of flavonoids. In: Andersen ØM, Markham KR (eds) FLAVONOIDS - chemistry, biochemistry and applications. CRC Press, Boca Raton

Meltzer H, Malterud K (1997) Can dietary flavonoids influence the development of coronary heart disease? Scand J Nutr 41:50–57

Minhalma M, Pinho MN (2001) Tannic-membrane interactions on ultrafiltration of cork processing wastewaters. Sep Purif Technol 22:479–488

Miranda I, Gominho J, Pereira H (2012) Incorporation of bark and tops in *Eucalyptus globulus* wood pulping. Bioresources 7:4350–4361

Miranda I, Gominho J, Mirra I, Pereira H (2013) Fractioning and chemical characterization of barks of *Betula pendula* and *Eucalyptus globulus*. Ind Crop Prod 41:299–305

Mota I (2011) Extracção de base aquosa de compostos polares da casca de *Eucalyptus globulus* na perspectiva da sua recuperação. Master thesis, Faculty of Engineering of University of Porto

Mota MIF, Pinto PCR, Novo CC, Sousa GDA, Neves Guerreiro ORF, Guerra ACR, Duarte MFP, Rodrigues AE (2013) *Eucalyptus globulus* bark as a source of polyphenolic compounds with biological activity. O Papel 74:57–64

Mota I et al (2012) Extraction of polyphenolic compounds from *Eucalyptus globulus* bark: process optimization and screening for biological activity. Ind Eng Chem Res 51:6991–7000

Moure A et al (2001) Natural Antioxidants from Residual Sources. Food Chem 72:145–171

Mukhatar H, Das M, Khan W, Wang Z, Bik D, Bickers D (1988) Exceptional activity of tannic acid among naturally occurring plant phenols in protecting against 7,12-dimethylbenz(α) anthracene-, benzo(α)pyrene-, 3-methylcholanthrene-, and N-methyl-N-nitrosourea-induced skin tumorigenesis in mice. Can Res 48:2361–2365

Neiva DM, Gominho J, Pereira H (2014) Modeling and optimization of *Eucalyptus globulus* bark and wood delignification using response surface methodology BioRes 9:2907–2921

Pinto PCR, Sousa G, Crispim F, Silvestre AJD, Neto CP (2013) *Eucalyptus globulus* bark as source of tannin extracts for application in leather industry. ACS Sustain Chem Eng 1:950–955

Pinto PCR, Mota IF, Loureiro JM, Rodrigues AE (2014a) Membrane performance and application of ultrafiltration and nanofiltration to ethanol/water extract of *Eucalyptus* bark. Sep Purif Technol 132:234–243

Pinto PR, Mota IF, Pereira CM, Ribeiro AM, Loureiro JM, Rodrigues AE (2017) Separation and recovery of polyphenols and carbohydrates from *Eucalyptus* bark extract by ultrafiltration/ diafiltration and adsorption processes. Sep Purif Technol 183:96–105

Pizzi A (2008) Tannins: major sources, properties and applications. In: Belgacem MN, Gandini A (eds) Monomers, polymers and composites from renewable resources, 1st edn. Elsevier, Oxford, pp 179–199

Rice-Evans C, Miller N, Paganga G (1996) Structure-Antioxidant activity relationships of flavonoids and phenolic acids. Free Radic Biol Med 20:933–956

Rice-Evans C, Miller N, Paganga G (1997) Antioxidant properties of phenolic compounds. Trends Plant Sci 2:152–159

Royer M, Prado M, García-Pérez ME, Diouf PN, Stevanovic T (2013) Study of nutraceutical, nutricosmetics and cosmeceutical potentials of polyphenolic bark extracts from Canadian forest species. Pharma Nutr 1:158–167

Russo C (2007) A new membrane process for the selective fractionation and total recovery of polyphenols, water and organic substances from vegetation waters (VW). J Membr Sci 288:239–246

Santamaría B, Salazar G, Beltrán S, Cabezas JL (2002) Membrane sequences for fractionation of polyphenolic extracts from defatted milled grape seeds. Desalination 148:103–109

Santos OS, Freire C, Domingues MR, Silvestre A, Neto P (2011) Characterization of phenolic components in polar extracts of *Eucalyptus globulus* Labill. Bark by high-performance liquid chromatography–mass spectrometry. J Agric Food Chem 59:9386–9393

Santos SAO, Villaverde JJ, Silva CM, Neto CP, Silvestre AJD (2012) Supercritical fluid extraction of phenolic compounds from *Eucalyptus globulus* Labill bark. J Supercrit Fluids 71:71–79

Song L (1998) Flux decline in crossflow microfiltration and ultrafiltration: mechanisms and modeling of membrane fouling. J Membr Sci 139:183–200

Stevanovic T, Diouf PN, Garcia-Perez ME (2009) Bioactive polyphenols from healthy diets and forest biomass. Curr Nutr Food Sci 5:264–295

Takuo O (2005) Systematics and health effects of chemically distinct tannins in medicinal plants. Phytochemistry 66:2012–2031

Thaipong K, Boonprakob U, Crosby K, Cisneros-Zevallos L, Byrne D (2006) Comparison of ABTS, DPPH, FRAP, and ORAC assays for estimating antioxidant activity from guava fruit extracts. J Food Compos Anal 19:669–675

Thorstensen TC (1993) Practical leather technology, 4th edn. Krieger Publishing Company, Florida

Vattem DA, Shetty K (2005) Biological functionality of ellagic acid: a review. J Food Biochem 29:234–266

Vázquez G, Fontenla E, Santos J, Freire MS, González-Álvarez J, Antorrena G (2008) Antioxidant activity and phenolic content of chestnut (*Castanea sativa*) shell and Eucalyptus (*Eucalyptus globulus*) bark extracts. Ind Crop Prod 28:279–285

Vázquez G, González-Alvarez J, Santos J, Freire MS, Antorrena G (2009) Evaluation of potential applications for chestnut (*Castanea sativa*) shell and eucalyptus (*Eucalyptus globulus*) bark extracts. Ind Crop Prod 29:364–370

Vázquez G, Santos J, Freire M, Antorrena G, González-Álvarez J (2012) Extraction of antioxidants from Eucalyptus (*Eucalyptus globulus*) bark. Wood Sci Technol 46:443–457

Velioglu YS, Mazza G, Gao L, Oomah BD (1998) Antioxidant Activity and Total Phenolics in Selected Fruits, Vegetables, and Grain Products. J Agric Food Chem 46:4113–4117

Wilcox LJ, Borradaile NM, Huff MW (1999) Antiatherogenic properties of naringenin, a citrus flavonoid. Cardiovasc Drug Rev 17:160–178

Yamaguchi F, Yoshimura Y, Nakazawa H, Ariga T (1999) Free radical scavenging activity of grape seed extract and antioxidants by electron spin resonance spectrometry in an H_2O_2/NaOH/DMSO system. J Agric Food Chem 47:2544–2548

Yazaki Y, Hillis WE (1976) Polyphenols of *Eucalyptus globulus*, *E. regnans* and *E. deglupta*. Phytochemistry 15:1180–1182

Yun B-S, Lee I-K, Kim J-P, Chung S-H, Shim G-S, Yoo I-D (2000) Lipid peroxidation inhibitory activity of some constituents isolated from the stem bark of *Eucalyptus globulus*. Arch Pharm Res 23:147

Zillich OV, Schweiggert-Weisz U, Eisner P, Kerscher M (2015) Polyphenols as active ingredients for cosmetic products. Int J Cosmet Sci 37:455–464

Chapter 5
Conclusions and Future Perspectives

Abstract Lignin valorization is nowadays an important research topic, both from academic and industrial point of view. This book reviews some key developments in the field, with an important contribution from the research conducted at LSRE-LCM, since more than 30 years. Pulp mills as biorefineries, integrated process for vanillin and syringaldehyde production from kraft lignin, polyurethanes from recovered and depolymerized lignins, and polyphenols from bark of eucalyptus globulus are the treated topics, which evidence the importance to approach lignin valorization from an integrated point of view, without neglecting the actual context of the pulp industry. Thus, lignin isolation from black liquor to be converted into chemicals (vanillin and syringaldehyde) and polyurethanes, and the integration of bark eucalyptus globulus valorization, contribute to a more sustainable approach and an enlarged concept of circular bio(economy).

Keywords Lignin valorization · Dicarboxylic acids · Polyurethanes · Purification and separation

In the actual context of lignocellulosic biorefineries' consolidation, and their search for sustainability and competitiveness, the effective valorization of lignin becomes a key issue. In fact, the complex macromolecular structure of lignin opens a wide range of possibilities that include fine chemicals, power/fuel, and macromolecules. Nevertheless, the constraints to full implement such strategies are various, demanding both fundamental and applied research, together with the investment in technological developments. In particular, the involvement of industrial players is decisive to make viable the scale-up of technologies for lignin extraction and applications development. An important point is the achievement of standardized raw lignins, crucial to assure their wide industrial use. This was the context that motivated the research behind the present book, which started almost 30 years ago in the LSRE-LCM group and continues to the present days.

From the emerging concept of biorefinery to the recognition of the pulp industry as a leading industrial sector in the field, the extension of the circular bio(economy), which started with pulp production (cellulose) and energy recovery (burning of the black liquor), was enlarged to include the proposal of integrating more rational strategies of lignin valorization (Galkin and Samec 2016; Rinaldi et al. 2016). This scheme considers, primarily, the isolation of lignin from the black liquor and their subsequent use in the obtainment of new chemical building blocks for diverse industrial uses (lignin platform). In this way, presently, the pulp industry is moving to a context where pulp production coexists with the sugar and lignin platforms, both important to support a new generation of biochemicals, biofuels, and advanced materials. Also, in the context of the pulp industry, it becomes increasingly interesting to explore directly the use of the pulping liquors as source of high-added-value chemicals and evaluate their economic advantages versus the direct use of lignin. Moreover, the full integration of all side streams of the pulp industry imposes bark valorization, in this work exemplified with the study of polyphenolic compounds extracted from *E. globulus* bark (polar fraction). In fact, these bioactive compounds have potential to be applied in different industrial uses (e.g., cosmetics, pharmaceutical, and leather industry, among others).

In the field of lignin valorization, the conversion of lignin from different origins into functionalized phenolic monomers was one of the topics with higher research investment in the past decades. This includes the use of lignin as a source of syringaldehyde and vanillin using environmental friendly processes such as the oxidation with O_2, which was the main focus of our research group, where oxidative depolymerization of lignins and black liquors was used for this end. In addition to the study of the reaction conditions favoring products yield and their respective kinetics, downstream processes to treat the obtained oxidized lignin mixture (e.g., membrane separation and chromatographic processes) were also proposed. The objective was to develop a feasible productive process, which includes production at batch and continuous mode, together with final separation and purification steps. We are convinced that cyclic adsorption processes using simulated moving bed technologies will be adequate to separate families of chemicals (acids, aldehydes, ketones) resulting from lignin oxidation (Gomes et al. 2018). Moreover, and in order to close the productive cycle, the proposal to use the residual depolymerized lignin in the synthesis of polyurethanes was also anticipated. The study comprised the use of the depolymerized lignins obtained as the by-product of the oxidation process to obtain vanillin and syringaldehyde to produce biopolyols by oxypropylation that were tested to produce rigid polyurethane foams.

Nowadays, lignin is still underutilized as a feedstock, and through depolymerization different aromatic building blocks can be produced (benzene, toluene, and xylenes) (Upton and Kasko 2016). Recent review papers on lignin depolymerization toward value-added chemicals can be found (Chen and Wan 2017; Demesa et al. 2015, 2017; Li et al. 2015; Upton and Kasko 2016). These works focus on conventional strategies but also give indications concerning new opportunities. Namely, there is a growing interest to obtain dicarboxylic acid from lignin (Cronin et al. 2017; Ma et al. 2014; Zeng et al. 2015). The process involves an oxidative cleavage of the aromatic ring. The 1,4-diacids (including succinic, fumaric, maleic, and malic acids) were selected as one of the 12 building blocks that will be the base of the chemical production derived from biomass (Werpy and Petersen 2004). Still a better understanding of the structural characterization of lignin and on the prediction of the maximum yield on target monomeric products is required (Evstigneyev 2018; Evstigneyev et al. 2017; Hu et al. 2016), namely, through the assess to alternative technologies supported by some kind of radar tool. Selective cleavage of alkyl aryl ether interunit bonds, mostly of the β-O-4 type, yielding monomeric aromatic products is one of the most promising avenues in this development. Evstigneyev (2018) suggests that the maximum yield of monomeric aromatic products from lignin degradation is $Y = (A-C)/(A + C)$ where A is a sum of α-O-4 and β-O-4 (n/100 PPU), and C is the sum of 5–5 and 4-O-5 bonds (n/100 PPU).

Currently, different approaches to generate added-value products from lignin can be found, and among them polymeric materials are assuming an important role (Duval and Lawoko 2014; Gandini and Lacerda 2015; Matsushita 2015). Lignin valorization in polyurethanes follows two main strategies: the use as such, i.e., without chemical modification, and the use after chemical modification (e.g., oxypropylation). Different lignin-based polyurethane systems are under investigation. These include rigid and flexible polyurethane foams, lignin-based composites, and thermoplastic polyurethanes (obtained from lignin as such, lignin-based polyols, depolymerized and fractionated lignins). Other topics include polyurethanes compounding with lignin to enhance biodegradability. Among the proposed strategies, the use of depolymerized lignins (directly or after oxypropylation), and of fractionated lignins, is gaining a rising interest, highlighting the advantages of using fractions with more controlled properties, instead of the whole lignin.

Facing to the actual context, lignin is consolidating its role in the field of chemicals and polymeric materials, which are gaining importance relatively to its use in fuel/energy applications. The market is expected to benefit from major industrial manufacturers' investments in R&D, such as in the development of improved technologies for lignin extraction and development of new products and applications.

References

Chen Z, Wan C (2017) Biological valorization strategies for converting lignin into fuels and chemicals. Renew Sust Energ Rev 73:610–621

Cronin DJ, Zhang X, Bartley J, Doherty WOS (2017) Lignin Depolymerization to Dicarboxylic acids with sodium Percarbonate. ACS Sustain Chem Eng 5:6253–6260

Demesa AG, Laari A, Turunen I, Sillanpää M (2015) Alkaline partial wet oxidation of lignin for the production of carboxylic acids. Chem Eng Technol 38:2270–2278

Demesa A, Laari A, Sillanpää M, Koiranen T (2017) Valorization of lignin by partial wet oxidation using sustainable Heteropoly acid catalysts. Molecules 22:1625

Duval A, Lawoko M (2014) A review on lignin-based polymeric, micro- and nano-structured materials. React Funct Polym 85:78–96

Evstigneyev EI (2018) Selective depolymerization of lignin: assessment of the yield of monomeric products. J Wood Chem Technol, Accepted

Evstigneyev EI, Kalugina AV, Ivanov AY, Vasilyev AV (2017) Contents of α-O-4 and β-O-4 bonds in native lignin and isolated lignin preparations. J Wood Chem Technol 37:294–306

Galkin MV, Samec JSM (2016) Lignin valorization through catalytic lignocellulose fractionation: a fundamental platform for the future biorefinery. ChemSusChem 9:1544–1558

Gandini A, Lacerda TM (2015) From monomers to polymers from renewable resources: recent advances. Prog Polym Sci 48:1–39

Gomes ED, Mota MI, Rodrigues AE (2018) Fractionation of acids, ketones and aldehydes from alkaline lignin oxidation solution with SP700 resin. Sep Purif Technol 194:256–264

Hu Z, Du X, Liu J, H-m C, Jameel H (2016) Structural characterization of pine Kraft lignin: bio-choice lignin vs indulin AT. J Wood Chem Technol 36:432–446

Li C, Zhao X, Wang A, Huber GW, Zhang T (2015) Catalytic transformation of lignin for the production of chemicals and fuels. Chem Rev 115:11559–11624

Ma R, Guo M, Zhang X (2014) Selective conversion of biorefinery lignin into dicarboxylic acids. ChemSusChem 7:412–415

Matsushita Y (2015) Conversion of technical lignins to functional materials with retained polymeric properties. J Wood Sci 61:230–250

Rinaldi R, Jastrzebski R, Clough MT, Ralph J, Kennema M, Bruijnincx PCA, Weckhuysen BM (2016) Paving the way for lignin valorisation: recent advances in bioengineering, biorefining and catalysis. Angew Chem Int Ed 55:8164–8215

Upton BM, Kasko AM (2016) Strategies for the conversion of lignin to high-value polymeric materials: review and perspective. Chem Rev 116:2275–2306

Werpy T, Petersen G (2004) Top value added chemicals from biomass: volume i - results of screening for potential candidates from sugars and synthesis gas. National Renewable Energy Laboratory (NREL), Golden

Zeng J, Yoo CG, Wang F, Pan X, Vermerris W, Tong Z (2015) Biomimetic Fenton-catalyzed lignin depolymerization to high-value aromatics and dicarboxylic acids. ChemSusChem 8:861–871

Index

© Springer Nature Switzerland AG 2018
A. E. Rodrigues et al., *An Integrated Approach for Added-Value Products from
Lignocellulosic Biorefineries*, https://doi.org/10.1007/978-3-319-99313-3

Printed in the United States
By Bookmasters